Objectivity, Invariance, and Convention

Objectivity, Invariance, and Convention

Symmetry in Physical Science

Talal A. Debs
Michael L. G. Redhead

Harvard University Press
Cambridge, Massachusetts, and London, England | 2007

Printed in the United States of America

Library of Congress Cataloging-in-Publication Data

Debs, Talal A.
Objectivity, invariance, and convention : symmetry in physical science /
Talal A. Debs and Michael L. G. Redhead.
p. cm.
Includes bibliographical references and index.
ISBN-13: 978-0-674-02298-0 (cloth : alk. paper)
ISBN-10: 0-674-02298-x (cloth : alk. paper)
1. Symmetry (Physics). 2. Simultaneity (Physics). 3. Objectivity.
4. Quantum theory. I. Redhead, Michael L. G. II. Title.

QC174.17.S9 D43 2007
39.7/25 225 2006051785

For Kristine and Jennifer,
who encouraged us in so many ways

Contents

Preface

As its title suggests, this work deals with two increasingly discussed topics: the use and significance of symmetry, or invariance, and the question of objectivity in the sciences. In doing so, this book attempts to provide a new appraisal of the philosophical significance of symmetry in the physical sciences. Part of this is provided by the refutation of an old suspicion among physicists and philosophers that, as a general rule, the invariance of a scientific model implies its objectivity. In fact, it is only for a very limited class of cases that such a claim may be maintained, and even in these cases invariance appears to have as much to do with convention as with objectivity.

In considering the role of symmetry in physics, this book also attempts to shed light on the larger current debate concerning objectivity in the sciences. Briefly, if science provides accounts of the world, then it would seem that these accounts must be objective to be of any value. Yet, they must also make use of so many social conventions—specialized languages, standards of measure, and so on—that this initial objectivity would appear to be undermined. This apparent tension between objectivity and conventionality in science is evident even within the ostensibly hard-science realm of modern physics. Thus, what follows is also an attempt to articulate one way of understanding scientific accounts of reality without falling into the pitfalls of extreme objectivist or constructivist positions. Since versions of these extremes appear explicitly and implicitly in the popular media, in literature, in discussions of public policy and ethics as well as within the academic literature, the underlying theme dealt with here is relevant beyond the boundaries of science studies.

Into these debates, sometimes dubbed the "Science Wars" or simply the "crisis of objectivity," this book interjects the view that the physical sciences, though dependent on convention, may nevertheless produce objective representations of reality. In demonstrating this, one must consider both the social and formal dimensions of scientific representation, taking into account the human subjects among whom representation actually takes place. In this view, human subjects depend on conventional choices to resolve ambiguity in representation. Nevertheless, scientific representation may still be substantially objective.

Finally, this book also attempts to illustrate the philosophical perspectives advocated by reference to detailed case studies taken from both of the major branches of modern physics, relativity theory and quantum mechanics. These studies, it is hoped, will also be considered contributions in their own right toward a better understanding of the way science works.

The topical relevance of this book is underscored by two recent contributions to the philosophical literature. The first of these is *Invariances: The Structure of the Objective World,* the last book of the Harvard philosopher Robert Nozick. Nozick's book, like this one, considers the proposal that invariance is a means to objectivity. Whereas his book advocates and extends this point of view, this book seeks to refute it as a general claim. In spite of their contradictory conclusions in regard to invariance, these works are in fact united by the desire to contribute to resolving the canonical tension between the philosophical relativist and objectivist.

The second major book to address this topic recently is the work of Stanford philosopher Patrick Suppes. Suppes' volume, *Representation and Invariance of Scientific Structures,* by contrast with Nozick's, provides numerous technical examples of the use of group theory in physics. In three technical case studies, the current work also seeks to draw upon detailed examples from modern physics. In case there was any doubt, the sheer volume of examples discussed by Suppes stands as evidence for the methodological and philosophical significance of symmetry in modern physics.[1]

Although this manuscript was initially written without the benefit of either of these books, far from pre-empting it, Nozick and Suppes implicitly or explicitly call for work like that presented here. In particular, Nozick writes that his "philosophical bent is to open possibilities for consideration," and that his intent is to "clear a philosophical space in which the newly proposed views can breathe and grow" (Nozick, *Invariances,* 3). Thus, this book owes much to Nozick for clearing space and to Suppes for raising interest in this topic, even though it provides its own novel perspective on the actual significance of symmetry in physical science.

Acknowledgments

This book is a collaborative effort between Talal Debs and Michael Redhead. Debs is responsible for the overall framework of the project, and especially for focusing on the relationship between objectivity and symmetry, on the one hand, and objectivity and convention, on the other. He is also responsible for developing the main lines of argument, often using Redhead as a reliable sounding board; in the process, Redhead has provided key insights into the development of the argument at critical junctures. In addition to this crucial role, he has helped immeasurably in sharpening the focus of the work, and in anticipating the voices of potential critics. Finally, Redhead's significant body of work in this field has formed an indispensable part of the conceptual backdrop for this inquiry, and this should be evident from the fact that a number of his key papers are referenced at different points in the book.

In addition, the book benefited especially from the hothouse atmosphere of the Department of History and Philosophy of Science of Cambridge University, as well as from numerous distinguished colleagues elsewhere. Jeremy Butterfield, Peter Lipton, and Simon Schaffer provided challenging insights and constructive criticism each from a very different methodological vantage point. Harvey Brown, Dennis Dieks, and Steven French also took the time to discuss portions of this work at length and contributed in substantial ways to its completion. It was at the encouragement of Gerald Holton that Debs first seriously pursued tailoring this work into a book. Additionally, the book has benefited from conversations with a number of people, including Guido Bacciagaluppi, Katherine Brading, Atif Debs, Caroline Debs, Mary Fisher, Jan Hilgevoord, Erwyn van der Meer, Simon Saunders, Mauricio Suarez, and Paul Teller. Also indispensible was the practical and moral support of Debs's longtime mentors Owen Gingerich and Henry Rosovsky.

In practical terms, the financial support that made this work possible included funding provided through the Arnold Gerstenberg Fund, the Committee of Vice-Chancellors and Principals of the Universities of the United Kingdom, and the Cambridge Commonwealth Trust. In addition, Debs offers sincere thanks to Everett Mendelsohn and Allan Brandt for hosting him at Harvard University during a critical period for this work, and this book would not have been possible without the vision and encouragement of Michael Fisher at Harvard University Press. We are also grateful to Eleanor Knox for assistance with the index.

Finally, the authors would like to acknowledge the generous support of the Center for Philosophy of Natural and Social Science at the London School of Economics and Political Science, where Redhead is the Co-Director and Debs is a Research Associate.

Objectivity, Invariance, and Convention

Introduction

The term "symmetry" refers to several distinct but related notions. Figure I.1 evokes at least two of them. The image reflected from the sunglass lens is the so-called "Temple of Bacchus" at Baalbek, in the Bekaa valley of Lebanon. This remarkably preserved first-century structure stands as an example of the significance placed on symmetry in antiquity, as related to aesthetic form and proportion. Another type of symmetry is also evident in the reflected image of the temple on the surface of the lens: Reflection, as formalized within modern physical science, is a symmetry transformation that, accordingly, preserves certain aspects of the original.

In addition to the theme of symmetry in its various forms, the photo evokes several important features of scientific representation. There is an original, the temple, and a person behind the shade of the glass receiving this information, but, as is usually the case in modern physics, the reader of this book does not have direct access either to the original in the world or to the image, or model, as it interacts with a person's mind. Instead, one has access to that which exists in the middle, which has often been called "media"—images, structures, or models that are used along the way in the task of scientific representation. Philosophers stand in relation to science as the reader does to this photo, with direct access only to the grainy and distorted medium of representation, the reflected image of the temple. The question may be posed as to how the symmetries displayed by these media relate to the process of representation; it is this question that serves as a starting point for the following inquiry into scientific representation.

Indeed, physicists and philosophers have long claimed a special role for symmetry in the physical sciences. Symmetry in modern physics has been formalized in terms of invariance as defined within the mathematics of group theory. For many, the significance of such symmetry is found in the claim that, in the words of Hermann Weyl, "objectivity means invariance

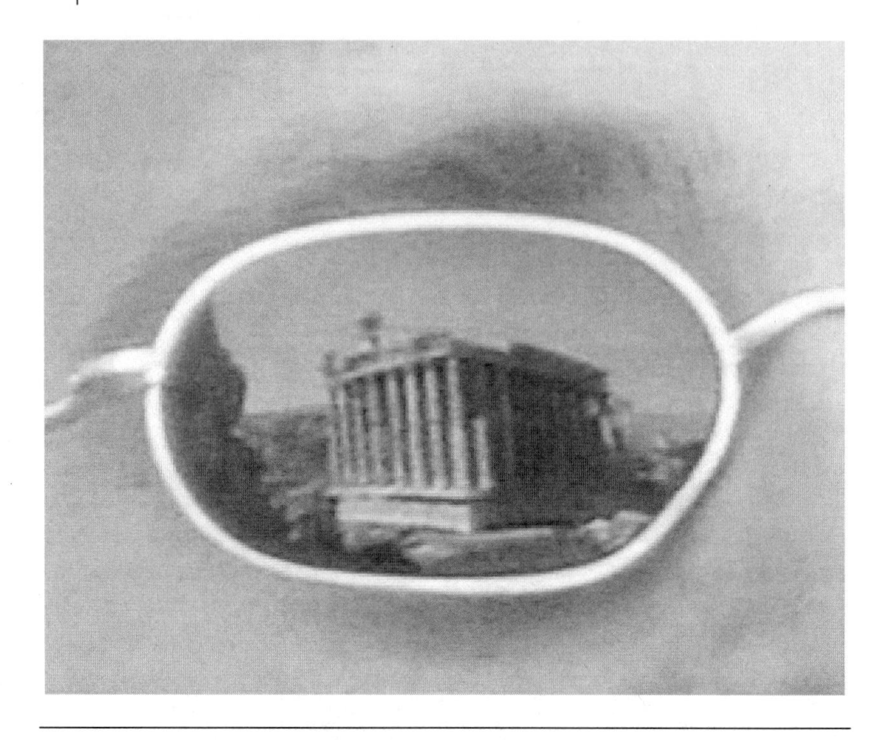

Figure I.1. Reflection of the ruins at Baalbek, Lebanon.

with respect to the group of automorphisms," or structure-preserving operations.[1] This approach to understanding the role of symmetry in scientific representation may be termed "invariantism."

It will be argued here that invariantism is not really adequate as a general account; the first step toward a new appraisal of symmetry. However, a limited form of perspectival invariantism does account for certain special cases in the physical sciences. In these cases, scientific models may meaningfully be called objective in a way that is linked to their invariance under specific symmetry transformations. Furthermore, it can be shown that the use of such invariant structures introduces a degree of representational flexibility or ambiguity, which must be managed through conventional choices; the second important step toward understanding the actual philosophical significance of symmetry in physics.

These observations will be illustrated through the consideration of case studies taken from the two predominant theoretical traditions within twentieth-century physics, relativity theory and quantum mechanics. The interpretative value of these case studies will, however, depend on one's initial perspective with respect to the tension between objectivity and conventionality in representation. Most observers agree that modern physical theory at least attempts to provide objective representations of reality. However, the claim that these representations are inescapably based on conventional choices has been taken by many as a denial of their objectivity. As such, objectivity and conventionality in representation have often been framed as polar opposites.

One's view on this purported tension will to an extent determine the significance placed on what follows. This is of course to be expected since most everyone works from an agenda of one form or another. Often this agenda is tacit and only comes to the surface through the sorts of problems considered and solutions provided; at other times it is more overt. The position advocated here bears significantly on two broad categories of such philosophical programs or agendas, each in a different way. As suggested above, these two categories involve those who support the objectivity of (at least some) representation in physics and those who argue for inescapable conventionality in representation; these categories might be called objectivist/anti-conventionalist and conventionalist, respectively.

In the context of the varying philosophical perspectives and agendas in the history and philosophy of science, the tension between conventionality and objectivity in representation is one of the most hotly contested. This might seem an overstated claim especially when one notes that in many ways conventionalist programs seem to have lost steam in recent years. Conventionalism in the tradition stretching back to Poincaré via Reichenbach and others has been considered, at least within the philosophy of physics, as essentially refuted for a number of years, due in part to David Malament's 1977 proof, which argued against conventionalist accounts of Einstein's special theory of relativity.[2] Similarly, in general philosophy of science, the debate that exercised the logical positivists and others between logical necessity and linguistic conventionality in physical laws has for the most part been left to lie fallow. In both cases there are still those few who periodically take up these conventionalist causes, but their voices sound more like the echoes of past disputes rather than the animated ones of current debate. However, this is not the whole story. The real battle between conventionality and objectivity, essentially dormant in the contexts mentioned above, has re-emerged in a slightly altered form and is taking place under different banners.

This re-emergence of tension between objectivist and conventionalist programs takes its form in two camps that have dominated the historical and philosophical study of science in recent decades. Although there remain significant skirmishes within these camps, the dispute between them, informally dubbed the "Science Wars," has caught the attention of many in other areas of the academy and in modern western culture as a whole. Of the many ways to attempt a distinction between these camps, one might make use of the broad labels "realist" and "social constructivist." Realists typically hold that empirically based scientific knowledge is at least a very good way to express the truth about the reality that humans inhabit, a truth that does not and must not depend on the vagaries of human activities and society. Social constructivists, on the other hand, would point out that scientific knowledge, like all knowledge, is expressed in language and is fundamentally a product of the cultures which create and use that language. One major source of friction between the two positions is that on some accounts the latter seems to imply that scientific knowledge is entirely relative to the culture that produced it in a way that rules out the possibility of objective scientific knowledge about reality.

The view taken here is that a correct understanding of representation in physics requires the recognition that both programs have points to make, in the following sense: although science is fundamentally cultural, it is a culture characterized by the activity of objectively representing reality. Moreover, a view of science that emphasizes the role of conventional choice need not be in conflict with a realist account of representation that allows for objectivity. One may maintain a cultural view of science and still be committed, as most realists are, to the existence of a single real ontology that humans inhabit.

This being the case, the conclusion argued for and supported by examples in this book will be significant to realist/objectivist and social constructivist/conventionalist programs in different ways. For the conventionalist it will demonstrate that an inescapable dependence on conventions in representation reaches well into the domain of the most technical theoretical traditions in the sciences. These instances are thus counter-examples to the suggestion that conventionality is only a feature of the sciences that display less theoretical rigor. For the objectivist/anti-conventionalist, the examples discussed here will stand as evidence against the claim that conventionality must undermine objectivity. After all, if it can be shown that under certain conditions invariance indicates objectivity, and at the same time introduces conventional ambiguity, then these cases should stand as evidence against such a polarized view. Thus, the intention is that the proposals made here will be of significance to both of these philosophical programs in the ways suggested. What follows is not then underpinned by a hybrid of realist and constructivist programs so much as an attempt to respect the fact that both have recognized important elements of an understanding of science that may, at least on the question of symmetry and its significance, not be contradictory.

This book is comprised of six chapters that address the question of the philosophical significance of symmetry in scientific representation. Chapter 1 builds up a view of representation in physics that takes account of both the formal and social dimensions of scientific representation. This is needed, ultimately because the claim that objectivity is linked to symmetry implicates both formal relations, through discussion of the symmetry of models, and social relations, through discussion of objectivity. The resulting simplified scheme focuses on three key relations—isomorphism, the representational relation, and the information relation. In addition, it is pointed out that an analogy to artistic performance is a suggestive shorthand for the way in which the social and formal dimensions of scientific representation interact.

Starting from this characterization of scientific representation, Chapter 2 goes on to lay further groundwork for a discussion of symmetry in physics. In particular, it introduces the notions of symmetry and invariance in greater detail and the ways in which the symmetries of scientific models actually introduce ambiguity in representation; ambiguity that must in turn be resolved by conventional choice.

Chapter 3 provides the essential features of the new appraisal of symmetry argued here. First of all, it characterizes the ways in which symmetry has seemed significant to philosophers of science, through its heuristic power, unifying virtues, and most of all as an indicator of objectivity. Second, it is suggested here that these themes can be brought together under an approach to symmetry which may be termed "invariantism." Although invariantist views are often proposed or assumed in the literature, they are not generally argued for. Thus, an argument is provided that might serve to justify the general claims of invariantism. Third, it is shown that invariantism as a general program may not be supportable. Fourth, it is shown that a limited form of perspectival invariantism does indeed serve to account for certain cases in modern physics for which invariance is both necessary and sufficient for a form of objectivity in representation. This fact accounts for much of the appeal that invariantism has had for physicists and philosophers alike. Finally, it is pointed out that even in these cases for which invariance implies objectivity, it also introduces a number of different sources of representational ambiguity, as initially suggested in Chapter 3. These must ultimately be resolved by conventional choice. Thus, though invariance does not generally imply objectivity, even where it does it also implicates convention.

Chapters 4, 5, and 6 each take up the challenge of demonstrating that this is the case with examples taken from modern physics. Chapter 4 picks up the theme of synchronization and simultaneity in special relativity, showing how the special status accorded to standard synchrony as defined by Einstein does indeed depend on conventional choices of one kind or another. Chapter 5 addresses the related and longstanding debate over the twin paradox. In particular, it presents a generalized scheme for understanding the ages of the twins. This scheme is based on the invariant quantity proper time, on the one hand, and the conventionality of standards of synchrony, on the other. Finally, Chapter 6 deals with the relatively technical problem within relativistic quantum mechanics of the localization of a single particle. Thus, as will be shown, one is faced with the choice of altering the relevant notion of objectivity or abandoning objective localization altogether.

1

Scientific Representation

It is often held that one of the great virtues of scientific knowledge is its objectivity. In subsequent chapters, this book addresses a specific claim that has been made about scientific objectivity, linking it to the invariant features of scientific models. These models are intended to represent reality. Though scientific practice certainly includes much that might be considered nonrepresentational, the construction of models to represent the world is a core activity. Thus, in order to address the question of the objectivity of scientific models, one must adopt some account of scientific representation.

There are two mutually interacting ways of analyzing scientific representation. One involves structures, models, and the mathematical and logical relations between them. The other involves the path of transmission, or how these structures come to have some representational interpretation; in other words the way in which they come to be taken by someone as representing a feature of the world through rhetoric and the mediation of networks of social agents. One might refer to these respectively as the formal and social dimensions of scientific representation. Moreover, because they interact in significant ways, it will be useful to adopt a combined account of scientific representation in what follows.

This combined account may seem unnecessary, especially if one holds the view that one or the other of these modes of analysis is foundational. If representation is only about formal relations, then it will seem an unwarranted burden to take into account the contingent social pathways through which they are established. If, alternatively, representation is only about social networks, rhetoric, and mediation, then why bother with the tedious (and on this view equally contingent) business of formal relations between structures?

There are good reasons, however, for adopting this combined account as a prelude to what follows. In fact, the claim alluded to above, that objectivity in physical science is related to the invariance features of its models, implicates both of these dimensions. In order to assess this claim, one must, on the one hand, discuss the invariance of specific features of a given model under a set of transformations; this is clearly within the realm of the formal dimension of scientific representation. On the other hand, one must assess what sense of the term of objectivity is intended. Most philosophical accounts of objectivity depend on some assessment of social relations in the sense that to claim that a fact is objective is often to claim that observers ought to agree about it. This second discussion falls within the social dimension of scientific representation. The fact that both dimensions of scientific representation are relevant to the claim that objectivity is linked to invariance will ultimately be helpful in demonstrating that invariance has as much to do with convention as with objectivity in the physical sciences.

This last observation also allows one to address a more general philosophical issue. Within the philosophical and sociological study of the sciences it is often assumed that there is a tension between the objectivity of science and its dependence on convention. This apparent tension often manifests itself in a question: Is scientific knowledge sometimes objective or always merely an agglomeration of socially negotiated conventions? The view taken here is that this is often merely an apparent tension. In order to make this point in greater detail, however, both the formal and social dimensions of scientific representation must be kept in the frame. This may be accomplished by a combined account of representation as outlined below.

Two Dimensions of Representation

As already suggested, there are two mutually interacting ways of analyzing scientific representation. The first involves models and logical relations; the second involves transmission of information via a social network and how these structures come to have some representational interpretation. A dual approach seeks to benefit from the insights of both dimensions of analysis. In presenting this combined account it is helpful to consider some relevant background.

Beginning at the beginning, one should note that the epistemic tension between objectivity and conventionality referred to above has an ontological cousin with an even more distinguished pedigree: the canonical question of the relationship between models and the "world out there," variously understood as some combination of mind-independent and culture-independent reality. It is some sense of the world out there that scientists generally claim to represent, regardless of whether they can produce a satisfactory account of how this is so.

This is related to the traditional philosophical problem of perception, which raises a very similar question, summarized in the schematic below (1): What is the nature of the relationship between perceptions in the mind of some agent, a person, with circumstances in the world?

World → Mind[1] (1)

This schematic captures what is likely the single most pervasive intuition about the nature of scientific inquiry: the theories and practical complexities of modern science may be thought of as extensions of our more or less direct perceptions of the world. This turns out to be a difficult intuition to justify.[2] The key term here is "direct." Asking this question about perception almost immediately recalls the tension between realism and idealism with regards to observer-independent reality, the world out there. It is a remarkable fact that this ancient question still stands in the background of much theorizing in a number of different fields, not least the philosophy, sociology, and history of late modern science.

The realist/idealist tension is notoriously difficult to resolve. A natural response to entering the swamp of lengthy and indecisive arguments in support of each of these ontological positions is to retreat to firmer ground. This is especially tempting if one is interested in how scientists actually do represent reality, assuming that they do succeed in just that. But what are the possible avenues of retreat? One hopeful starting point is to limit the conversation about the perceiver/scientist and external reality, and speak instead of "media," that which exists between them. If one accepts that in practice scientific representation is indirect, in the sense that it is mediated in some way through use of instruments, mathematical models, and scientists themselves, then one may adopt a new schematic that involves a refinement of (1), bracketing the metaphysical question of how world and mind connect:

World [Media] Mind (2)

If one is interested primarily in what goes on inside the square brackets, this is no longer the problem of perception, but instead a problem of understanding scientific representation.[3] More specifically, one may ask how it is that certain structures or models come to be taken by specific people as representations of reality.

Having taken this step, summarized in schematic (2) above, one is left with another dualistic tension over how to interpret the relationships among the instruments, models, and people that fit under the heading of "media" as it appears above. According to one approach, exemplified within science studies by social constructivism, scientific representation is fundamentally about social agents and the instrumentation or other apparatus used to mediate information across a social network. An opposing view might maintain that understanding scientific representation depends primarily on the formal analysis of structures; support for this approach may be found among the various proponents of structural realism.

For the social constructivist, firm ground is that of social relations. Why worry about perceiving the world "out there" when one can discuss the ways in which representations of the world are made by and for human beings? On this view representation is no longer about relating people's perceptions to the world, summarized in schematic (1), but instead it is about how social relations mediate scientific knowledge. A schematic of this position privileges these social relations:

World [. . . Person → Material Culture → Person . . .] Mind (3)

For the social constructivist, then, representation involves a chain of media that includes especially social agents and physical objects, often referred to as "material culture," since these objects have specialized socially relevant functions. In applying this to scientific practice there is often more than one scientist or piece of apparatus in this social chain of media along which information is transferred. While this account is clear on who is mediating scientific knowledge and how this happens, it is relatively uninterested in what, beyond social relations, determines the sorts of theories and models thus disseminated. Similarly, it is often felt that this approach is exclusively interested in the rhetorical aspect of scientific practice to the exclusion of possible nonrhetorical limitations on theory choice. Thus social constructivism is often saddled (fairly or unfairly) with the reputation of extreme relativism. To the extent that this emphasis on the social is pushed explicitly or implicitly to a metaphysical preference for social reality, this view transgresses the boundary established above in restricting the discussion to the media of scientific representation. While avoiding such ontological commitments, one may nevertheless consider it a significant observation that there is a social dimension of scientific representation, mediated by social agents and material culture in the way suggested above, summarized in schematic (3).

According to an alternative approach, however, firm ground is found in the realm of structure and formal relations, especially the structures that appear in scientific theories. This is exemplified particularly well by the various proponents of versions of structural realism. This formal approach represents the long-standing tradition of framing the problem of representation as being about matching structures to one another:

World [Idealized Original → Mathematical Model] Mind (4)

Here, the analog to the social chain of media above (see schematic (3)) includes primarily two structures, an idealization and a mathematical model, that are connected by formal relations. By contrast with the social chain of media, this formal chain of media does not mediate knowledge between people, but does mediate relations, isomorphism for instance, between structures.

There are two sorts of questions raised by this approach to scientific representation. First, what structural features suggest that a given model may be considered to be a representation of a physical object or system in the world? The structural realist may offer a number of suggestions in response to the first question, but in doing so would be venturing back into the realm of ontology.

Second, what makes any concrete structure, a painting or mathematical model, a representation of any other? In response to this question, the formal approach is by contrast unable to provide much of an answer. This is due to the fact that an explanation of why one structure is a model of another introduces an element of interpretation that goes beyond the formal analysis of structure. To illustrate this point, consider an example from the arts: A certain painting, a still life, may resemble a bowl of fruit, but there is nothing about the pattern of paint on the canvas, on the one hand, or the bowl of fruit, on the other, to indicate which is representing which. Although the relevance of resemblance to representation will be discussed in the following section, it is sufficient here to point out that it is the wider social context that allows one to identify the painting as the representation of the bowl of fruit. The analogous conclusion also holds with regard to scientific representation: Accounting for why a specific structure is taken as a model depends on both formal analysis of its features and an awareness of the social dimension.

Up to this point, one might have hoped that a combined account of scientific representation would decompose neatly along the lines of these two dimensions, social and formal, as distinct and noninteracting regimes, something like the so-called "context of discovery" and "context of justification." In reality, however, these dimensions interact in significant ways. Not only does the formal dimension depend on the social, as suggested above, the social depends on the formal in the sense that formal relations, like resemblance or isomorphism, often constrain the sorts of models that are accepted and propagated by scientists. This is evident when conventional choices, which are made within the social dimension, can be shown to depend on formal ambiguities in the models being applied to a given phenomenon. Such cases will be discussed in detail.

In summary, this combined account considers as relevant both the social and formal dimensions discussed above. In doing so, one must keep in mind that the media of scientific representation include both a social chain of media and a formal chain of media, along which distinct types of relations are mediated. Thus this view may be summarized by a combination of the schematics above:

[. . . Person → Material Culture → Person . . .]
World   Mind   (5)
[Idealized Original → Mathematical Model]

That these dimensions interact in significant ways is an important part of the approach taken here. In order to see this more clearly, more must be said about the social and formal relations that are established between social agents (scientists) on the one hand and structures (scientific models) on the other. To this end, a closer look at each of these two dimensions of scientific representation will be helpful.

The Social Dimension

The social aspect of scientific practice in general and representation in particular is a vast area of study. This field also crosses over into other sophisticated branches of inquiry, including sociology, psychology, cognitive science, and decision theory, to name but a few. This literature includes numerous approaches that one might use to characterize relevant social interactions between scientists. What is needed here, however, is a basic (and minimally complex) characterization of ways that scientists interact as social agents when they claim to represent the world with a given mathematical model.

The Information Relation

In practice, finding such a thing is not easy. There is a reason the aforementioned fields display such an array of possible approaches to the problem: social interaction between humans is complex to a degree that often defies analytical treatment. Mercifully, the issue of central interest here is considerably narrower. The discussion has already been delimited to an extent by the decision to focus on the social chain of media, described above. It consists of scientists and apparatus that intervene between an object in the world and its representation in the mind of a knowing subject. This representation is mediated, or transferred, down this chain from one social agent to the next. This is accomplished as one scientist communicates with another, using various tools. Thus, the social relation of greatest obvious significance here is one of sharing or transferring information.

To put some flesh on these bones, however, it may be pointed out that the transfer of information includes a strong rhetorical component. Indeed, a large part of what a scientist must do is to convince others that a given model really does represent some feature of reality in the world. Thus the direction in which information is transferred is revealed in the (social) relationship between scientists and their specialized scientific audience. If one represents the social chain of media schematically, moving from phenomena in the world on the left to a representation in the mind on the right, as in (3) above, then information is mediated by this chain from left to right. Each person in this chain, except perhaps the leftmost, unless a circular arrangement is contemplated, will have their turn at being the audience and target of the rhetorical activities of the person to their left. They will also have a turn at presenting that information to the next person or audience down the line. If one refers to the latter role of presenting information as being an "actor," for reasons which will be explained shortly, then a revised schematic of the social chain of media looks like this:

World [. . . → Audience/Actor → Audience/Actor → . . .] Mind (6)

This time the arrows are given an explicit interpretation as indicating the flow of information. Thus, the social mediation of a scientific representation may in principle be broken down into a series of binary interpersonal relationships, established with the purpose of transferring this information. For the sake of simplicity, this schematic de-emphasizes the material culture and apparatus, which are properly speaking part of the social chain of media. However, that these features should be implicit in that apparatus is a core aspect of the rhetorical activity depicted here. This is of course a gross oversimplification of an immensely complex activity, but it captures a central feature of the social aspect of scientific representation in a straightforward manner.

One should at the outset settle an obvious objection to the way this "information relation" has been introduced. In reality scientific practice is distributed across a complex social network, not along linear chains of binary relations. The simplification of a two-place information relation nevertheless accommodates this reality, as larger scientific audiences or groups of scientists working together may be understood in terms of the combination of a number of these two-person relations. The social mediation of a specific element of information will take place across some path in this network, each of which may be decomposed into a series of two-place information relations.

It might also be suggested that in some cases the information relation should hold between less than two people because it should be perfectly possible for a single person to represent reality to themselves. The notion of social mediation of information from one person to him or herself is at best a degenerate case of "social" mediation. In addition, one might resort to Wittgenstein's argument ruling out the possibility of private languages.[4] In essence, representing reality to oneself might require the use of language that can only take its meaning from a wider social context.

Thus the information relation as introduced above is a perfectly coherent albeit simplistic means of capturing a complex phenomenon. For the purposes of the discussion that follows, the social dimension of representation will be characterized by social mediation of information. An analysis of this dimension of representation will therefore depend on the individuals between whom the information relation holds. It is these individuals, along with the apparatus they use, who constitute the social chain of media.

The Formal Dimension

As previously suggested, the information thus socially mediated is usually expressed in terms of formal relations. It will be crucial to consider these relations in detail as aspects of the formal dimension of scientific representation, which may be conceptualized in terms of structures and the relationships between them.[5] The most obvious formal relation relevant to representation, resemblance, has already been mentioned. In fact, the notion of resemblance is an issue that dominates the philosophical discussion of representation in the arts.[6] One might similarly wonder to what extent scientific representation may be analyzed in terms of the notion of resemblance. It has often been argued that representation, unlike resemblance, is neither reflexive nor symmetric.[7] If a depiction, *D*, represents an original, *O*, then *O* generally does not represent *D*, nor does *D* represent itself, even if *D* and *O* resemble each other closely. It is thus argued that resemblance is not a good notion with which to analyze representation.

Scientific representation, nevertheless, relies heavily on the notion of resemblance, as expressed in terms of isomorphism between structures. The objection raised above is based on the ordinary linguistic meaning of "*D* represents *O*," and the specialized notion of scientific representation need not be constrained by such general considerations. In fact, more than one philosopher of science has despaired of success when faced with the problem of putting forward an acceptable general account of the sort of representation that takes place in the sciences.[8] Part of the task here is to develop a useful understanding of the way in which scientific representation actually takes place in the context of modern physics. This need not be constrained by the traditional arguments that divorce artistic representation from resemblance.

In practice, resemblance in the form of similarity between structures is basic to a formal analysis of representation in the physical sciences. In general, a physicist sets out to represent some particular physical system, *W*, in the world. A scientific representation of *W* depends on two more structures, an idealized conceptual model, *O*, and a mathematical model, *M*. Structures *O* and *M* constitute the formal chain of media, while *W* stands for some feature of the world, revising (4) above:

$$\text{World } (W)\ [O \leftrightarrows M]\ \text{Mind} \tag{7}$$

There are two types of relation that exist between these media. The first, indicated by the double arrow, involves the notion of structural similarity, especially isomorphism. The second, indicated by the single arrow, is conceptual and has to do with the sense in which structures are understood to represent one another. This second representational relation may indicate that a given structure is either the model of an original or the token of a type. Before addressing the much-debated issue of isomorphism and representation, the representational relation will be considered in more detail.

The Representational Relation and Structure

A representational relation may be said to exist between structures in one of two senses or modes. First of all, one may conceive of representation in terms of tokens and types, according to which a token structure, *t*, "stands for" and thereby represents a category of structures or a type, *T*. Alternatively representation may involve a model, *M*, which is intended to carry or express some salient features of an original structure, *O*. In the framework introduced above, *O* is an idealized model of some physical system, *W*, in the world. These two modes may operate together and overlay one another. For instance, a standard ball-and-stick model of the hemoglobin molecule may stand as a token for the type of molecule known as hemoglobin. At the same time this structure may be understood as modeling the features of a single such molecule as it undergoes the biochemical process of binding oxygen.

The distinction made between token/type and model/original modes of representation is useful in clarifying a counterintuitive usage of the term "representation" often found in the physics literature. In this literature it is sometimes asserted that a given theoretical structure is a representation of a physical one, and at other times asserted that a given *physical* structure is a representation of some *theoretical* one. Since physics attempts to construct theories that represent physical reality, this second situation might appear to be the reverse of what physicists would like to accomplish. In reality, this situation may be understood consistently as an example of the overlaying of modes of representation: a physical structure, *W*, may be represented by idealized model *O* and a mathematical model, *M*, which is at the same time a token, *t*, for a type, *T*. In this way, one might speak of model *M* as representing *O*, *and* of *O* as standing (as a token) for the type *T*.

In this scenario, certain structures (the type, *T*, for example) are more abstract than others. In order to clarify the distinction between abstract and concrete structures one must first take a step back and define more carefully what is meant by the term structure in the first place. Indeed, a discussion of the formal dimension of representation, as it involves the relationship between structures (as originals, models, tokens, or types) relies on a formal understanding of structure. A structure may be defined as an entity (or object) comprised of other sub-entities that bear specific relationships to one another. In this most general sense, examples of structure might be found in physical objects, mathematical or geometrical entities, and even paintings or narrative accounts. There are two useful ways of distinguishing between structures. These include the distinctions between concrete and abstract structures, and between algebraic and nonalgebraic structures.[9]

In addition, a third distinction is often made that contrasts physical against mathematical structures. This is based in turn on the traditional division over the ontological status of mathematical and physical objects respectively, and is not particularly useful in the context of this discussion of representation. That is to say that the use of mathematics in scientific representation is neutral with respect to the ontological status of mathematical objects. Recalling the traditional positions, a platonist view of mathematical objects as existing independently of the physical world that mathematicians inhabit may be contrasted to a view that denies their independent existence. This independent existence is often thought of in terms of platonic forms beyond the direct access of physical reality. Since one is interested here in the *use* of mathematical theory in representation, an activity that takes place entirely within the same physical reality which the physicist or mathematician inhabits, then the platonist claim is, strictly speaking, outside the scope of this discussion.

However, it remains a fact of theoretical practice that physicists very often act as if mathematical objects and structures do exist in something like a platonic independent sense. The most significant implication of this predilection of physicists is that within representational practice they treat mathematical and physical structures very similarly. In order to allow for this, one may adopt a view of mathematical objects that treats them as if they were independently existing objects without actually taking a position on their metaphysical status. Such a view has been proposed by Stewart Shapiro as "working realism," or alternatively by Michael Resnik as "methodological platonism," which treats mathematical and physical structures "on a par."[10] For the purposes of this discussion, these structures are on a par with one another in the sense that physicists may equally well make use of mathematical structures/objects and physical structures/objects in representing reality. Furthermore, conceived as on a par with one another, a fixed direction of representation need not be implied; a physical structure may be used to represent a mathematical one just as well as the reverse.[11]

A more useful distinction is that between concrete and abstract structures. A concrete structure may be defined as any particular set of elements plus the relations between them. Thus a collection of physical objects in a certain arrangement, for instance the triangle formed by the summits of three alpine peaks, may be considered a concrete structure. Similarly, a particular mathematical structure, the triangle defined by three given points in a Euclidean plane, may be considered a concrete structure. The use of the term "concrete" to refer to mathematical structures as well as physical ones may seem counterintuitive at first, especially since the term, in general linguistic use, usually indicates a certain physical tangibility. However, in light of the methodological platonist view adopted above, mathematical and physical structures may be treated as if they were on a par with one another. It is then the particularity of a structure that makes it concrete.

If concrete structures are particular, then abstract structures are intended to be, in some sense, universal. One may thus attempt to arrive at a structure by abstraction from particular concrete structures. The result of such abstraction, an abstract structure, may then be defined as a universal structure that shares certain features with particular concrete structures. In the case of the concrete triangles suggested above, those defined by alpine summits and Euclidean points respectively, an abstract triangle might be said to share the general feature of having three noncollinear elements. Thus these concrete structures might be considered as particular instances of the universal abstract structure commonly called a triangle.

This sort of abstraction might be interpreted either as dependent on platonic forms or alternatively in a nominalist sense. First of all, an abstract structure may be conceived as a platonic form, existing independently of the concrete structures that instantiate it. On this view, an abstract structure is the ideal universal form of any particular combination of elements and relations that possess certain structural relations. Secondly and alternatively, an abstract structure may be understood (in nominalist terms) as simply the set of all concrete structures that possess those structural relations. The platonic interpretation of abstract structure involves a metaphysical claim about the existence of independent idealized objects, the abstract structure interpreted as a form. This presents a difficulty since it is just this metaphysical claim that methodological platonism seeks to avoid.

One can avoid this difficulty, however, by opting for a nominalist interpretation of abstract structures. On this view, an abstract triangle would be understood as a collection of concrete triangles. This collection would include physical and mathematical structures, all of which exist as concrete structures in the sense introduced already. Accordingly, the abstract structure is simply the structure (elements and their relations) that is shared (and hence defined) by a collection of concrete structures. To see how this avoids the implication of independent platonic objects, consider the following illustrative case. Taking once more the abstract structure of a triangle that is associated with a set of concrete structures which share the structural feature of having three noncollinear elements, one may gradually reduce this set to a single concrete structure by removing particular concrete structures from consideration. One may thus end up with a set that only contains one concrete structure. Once this has been done then talk of sharing features among the members of the set becomes trivial and vacuous. To say that this concrete structure shares features with itself may be formally valid, but those features include much more structure than that of a triangle (in fact all of its structural features). In this case then there is no longer an abstract structure to speak of; the abstract structure simply does not exist independently of a collection of concrete structures. The key point of this example is that the abstract structure does not have the independent existence that one would expect if it were a platonic form. Thus this interpretation of abstract structure, as that shared between concrete structures, need not imply platonic metaphysical claims.

As previously suggested, the distinction between concrete and abstract structures effectively accommodates the two modes of representational practice that have been introduced as token/type and model/original representation. In the latter, model/original, form of representation, a model, *M,* models the features of an original, *O.* Consider again a simple ball-and-stick model of hemoglobin used to represent a single such molecule. In this case, both *M* and *O* are particular structures and hence concrete. One may recall that this model/original mode of representation often coexists with the other token/type mode, so that the same structure that is used as model *M* of the hemoglobin molecule may also stand as a token, *t,* for the type, *T,* of molecule known as hemoglobin. In this second instance, there is a move in emphasis from particular toward universal; this involves the sense of abstraction captured by what has been called abstract structure. This can be seen from the fact that an abstract structure depends on a collection of similar concrete structures, any member of which may be used as a token for the type defined by their shared features. Thus, model/original representation makes use of concrete structures and token/type representation makes use of abstract structures as well as concrete ones.

There is, additionally, a useful way of classifying abstract structures in terms of the distinction between algebraic and nonalgebraic structure recently made by Stewart Shapiro.[12] This distinction is drawn in the context of the ways in which sets of structures may be compared and categorized. This is similar to the situation one faces when dealing with abstract structures and the sets of concrete structures on which they are based. Adapting Shapiro's definition to this discussion, an algebraic abstract structure is one that is associated with a set of concrete structures which share certain features but are not entirely identical or, more precisely, isomorphic to one another.[13] Conversely, a nonalgebraic abstract structure is one that is based on a set of isomorphic concrete structures.

To illustrate the difference between algebraic and nonalgebraic structures, consider once more the case of triangles. A particular triangle, a concrete structure, may also stand as a token for an abstract structure with three noncollinear elements, an abstract triangle. This most general abstract triangle is based on the set of all concrete triangles, many of which may not be congruent with one another, hence one is dealing with an algebraic abstract structure. If, on the other hand, one restricts membership of the set of concrete triangles to those that are congruent, equilateral triangles for example, the abstract structure so specified would be nonalgebraic.

This distinction between algebraic and nonalgebraic abstract structures is especially applicable to this discussion of representation because it applies to the structure of groups within group theory, the mathematical basis for the use of symmetry concepts in modern physics. The structure of a group is in fact Shapiro's primary example of an algebraic structure; he points out that the concept of a group is a category not "unique up to isomorphism," noting that some groups are commutative and others not.[14] On the other hand a particular group (specified by a multiplication table) may be classified as a nonalgebraic abstract structure.

A final observation with regard to the notion of abstract structure is that it seems to invite one to ask the difficult question of whether relations may be said to exist without elements. If so, then a structure might refer to a network of relations without *relata*. In fact, this is really only a problem if one adopts the first, platonic, interpretation of abstract structure presented above. On this view of abstract structure as an idealized form, the form exists as independent of structures in the real world. One might try to argue from this that the abstract form is entirely characterized by relations independent from the particular elements between which they exist.

This would imply a very strange meaning of the term "relationship," which after all is usually held to exist between elements of one sort or another. However, one may avoid this uncomfortable situation by adopting an interpretation of abstract structure as a collection of concrete structures. According to this view, one may not speak of an abstract structure in terms of relations alone. If, as previously suggested, an abstract structure must always be instantiated by at least one concrete structure with its full complement of relations and elements, one may avoid the possible criticism that this discussion of structure in representation implies relations without *relata*.

Having introduced some of the definitions and distinctions relevant to the structures used in scientific representation, one may return to the notion of the representational relation between them. Although there are in general two senses in which one structure may be said to represent another, as the token of a type or the model of an original, what follows will focus on the latter. In practice, the use of scientific models is central to understanding representation in the physical sciences. In considering a physical system in the world, *W*, its conceptual model, *O*, and its mathematical representation, *M*, one may see both concrete and abstract structures as well as both modes of representation at work. Physical system *W* and idealized model *O* may be concrete structures or may be understood in an abstract sense as classes of physical systems or idealized models. These structures may in turn be represented by a concrete structure, mathematical model *M*, either as the model of an original (concrete conceptual model *O*) or the token of a type (defined by an abstract conceptual model *O*).

To illustrate how this might apply to physics, consider the scientific representation of X-radiation. Modern physics models this phenomenon with the concept of an electromagnetic wave. This wave-like behavior may be modeled mathematically through the use of orthogonal sinusoidal functions summarized by Maxwell's equations. In this case the physical system in the world, *W*, is X-radiation, and it is represented by conceptual model *O*, electromagnetic waves. These waves are in turn modeled by a mathematical model *M*, Maxwell's equations. There are two relations of special interest here, between *W* and *O*, and between *O* and *M*. Of the former relation one may ask the question, "How does one determine that X-radiation in the lab may be represented as a wave-like phenomenon?" Of the latter relation, one may ask the question, "How does a mathematical model, Maxwell's equations, represent a wave-like phenomenon?" What this example suggests is that characterizing scientific representation in terms of these three structures helps maintain an important distinction between two different aspects of scientific representation, already discussed at the start of this chapter in terms of the "problem of perception" and that of scientific representation. The former is the philosophical problem of how one connects a conceptual model *O* with the physical system *W* in the world; how does one gain sensory or cognitive access to the mind-independent world? The latter brackets this question and focuses instead on the media of scientific representation. From the perspective of the formal dimension of representation, this may be accomplished by studying the relationship between an idealized model, *O,* and its mathematical model, *M.*

Isomorphism and Representation

In addition to the representational relation between *M* and *O,* there remains the issue of similarity between these structures. By creating a mathematical model, *M,* of *O,* the physicist attempts to capture the features of the structure of *O.* Recalling the notion that a structure is characterized by its elements and the relations between them, the structural features of *M* and *O* depend on the same. In general, there will be some structural similarity between structures *M* and *O.* Thus, the notion of similarity here may be cashed out in terms of a homomorphism, mapping *O* to *M,* which preserves the structure of *O.* In an ideal case these structures will be similar in every possible regard. In this case, the homomorphism will be bijective (both onto and one-to-one), and this mapping is called an isomorphism. When *M* is isomorphic to *O* then one might say that the conceptual model *O* is structurally identical to its mathematical model *M;* i.e., they are the entirely the same in regard to their elements and the relations between them.

At the outset, one must emphasize that representation in physics is not generally intended to capture *all* the features of the original. In the case of model *M,* there is in general an attempt to represent accurately only a simplified, idealized version of the original physical structure in the world, *W.* A relation of strict isomorphism only exists between the model, *M,* and the idealization, *O,* not between the model and the world, *W.* The relation between the model and the world can be expressed, however, via the formal notion of *partial* isomorphism, as developed for example in the work of da Costa and French.[15]

Analyzing representation in terms of structures is often taken to imply that structural similarity is either necessary or sufficient for representation. That is to say, one might be tempted to believe that if *M* represents *O* then it is isomorphic to *O,* or alternatively that if *M* is isomorphic to *O,* then it represents *O*. Strictly speaking, neither is the case, and accounts that view scientific representation purely in terms of isomorphism have rightly come under attack for a variety of reasons.[16] One obvious objection derives from the asymmetry of the representational relation: model *M* cannot automatically represent *O* by virtue of their structural similarity, but is instead used by the physicist in such a way as to make the claim that *M* represents *O* and not that *O* represents *M*. That is to say that isomorphism between *M* and *O* underdetermines the direction of the representational relation between them. This is exactly analogous to the example above of a painting of a bowl of fruit. The most promising solution to this problem looks to the intentions of the physicists using model *M*. This is one place in which an understanding of the formal dimension of scientific representation depends on considerations taken from the social dimension.

This limitation of the role of structural similarity notwithstanding, it is reasonable to suggest that the existence of a homomorphic mapping from *O* to *M* is necessary but not sufficient for scientific representation. This suggestion derives from the simple fact that scientists apply mathematics to the task of representation with the intention of capturing structural similarities. If there were no homomorphic mapping from a conceptual idealization *O* to its mathematical model *M*, then *M* would in practice only be viewed as a scientific representation of *O* in the most trivial sense, if at all.

In many cases, *M* is taken to be isomorphic to *O*, although there are also often cases in which there is a homomorphic map from *O* to *M* that is onto but not one-to-one. In such cases, a certain amount of structure is either compressed into or left idle within the mathematical model *M*. To illustrate the latter possibility, consider the case in which mathematical model *M* lies in the larger mathematical structure, *M′*, in which it is embedded. The complement of *M* in *M′* represents what one of the authors has called "surplus structure."[17] To see what is meant by this, the three-structure scheme of representation (*W*, *O*, and *M*) utilized up to now may be expanded to include surplus structure (see Figure 1.1).

One may think of mathematical model *M* as a substructure of a larger mathematical structure, *M′*, which is needed for the construction of model *M*, but elements of which are not considered to represent elements of the idealized model, *O*. These might include, for example, complex numbers that appear as phase factors in quantum mechanical states; such numbers lead to predictions about real observable quantities, but are not themselves considered to refer to physically real quantities.

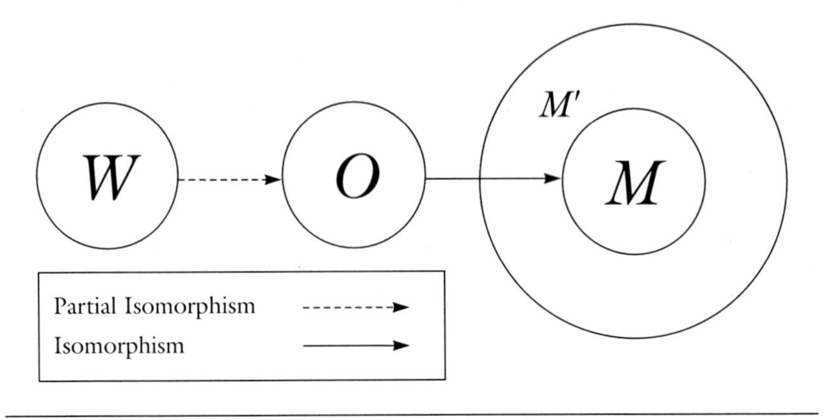

Figure 1.1. Formal relations between representational structures *W*, *O*, *M*, and *M′*.

The very fact that mathematical surplus structure is often in evidence raises an interesting philosophical question about how one really determines which features of the larger structure *M′* are to be considered as representing aspects of the physical system in the world. In other words, the boundary between *M* and its complement in *M′*, the surplus structure, is to some extent elastic. There are numerous examples from the history of physical sciences of elements of surplus structure eventually "graduating" to being considered as elements of the model *M*; these include notably the notion of energy, the sum of potential and kinetic energy, which eventually came to be considered a physical quantity in its own right.

The relations summarized in Figure 1.1, taken together with the representational relation between *M* and *O*, characterize the aspects of the formal dimension of scientific representation most relevant to the issues pursued here. Whereas the social dimension of representation facilitates social mediation of scientific knowledge, the formal is concerned with mediating these relations along a chain of media that in its greatest generality includes *O*, *M*, and *M′* (the surplus structure). In practice, analysis will usually be limited to the features of idealized conceptual model *O*, mathematical model *M*, and the relations between them.

A Combined Account

Thus far, two dimensions of scientific representation, social and formal, have been considered. It must now be re-emphasized that these two dimensions are not conceptually orthogonal to one another; in other words scientific representation does not decompose neatly into two non-interacting domains. In reality, they interact with one another and each places constraints upon the other. Formal relations without clarity as to what information they are intended to convey are no more representational than a rhetorical claim devoid of formal structural content. Understanding the interaction between the formal and social is therefore an important step in considering both dimensions of this combined account of scientific representation.

This may be attempted in a couple of ways. First of all, one may focus on the relations discussed up to this point; the overall picture that emerges places emphasis on three basic relations: an information relation between actor and audience in a social chain of media, isomorphism between structures in a formal chain of media, and a conceptual representational relation between a model and an original. One may thus point to several specific ways in which these relations interact with and depend upon one another. This is effectively an extension of the simplified accounts of the social and formal dimensions introduced above, and it is all that will be needed to take the discussion further in subsequent chapters.

In addition to this idealized combined account, however, there is a second approach that further illustrates the interaction between the social and formal dimensions of scientific representation. One may gain a more intuitive appreciation for the intimate relationship between formal and social in scientific representation by considering an analogy with artistic performance. Artistic performances depend equally on social relations with an audience and the formal aspects of artistic composition. This analogy is intended to provide a suggestive and intuitively appealing way to think about a combined account of scientific representation. It will not add more detail to the idealized combined account adopted here, but may for some be a helpful way of thinking about scientific representation.

Interaction between Key Relations

A combined account of scientific representation emphasizes three key relations: the information relation, the representational relation, and isomorphism. Two ways in which these relations interact have already been mentioned. First of all, as previously suggested, the direction of the formal representational relation is dependent on the social dimension of representation. Secondly, it was also claimed that the sort of information which social agents mediate includes formal content and is thereby subject to formal constraints. Each of these points may be pursued a little further. The former suggests something more general about representational ambiguity and convention. The latter is related to an additional sense in which formal and social dimensions interact as a result of the so-called "theory-ladenness" of experiment.

There are two additional notions that are central to understanding the interaction between social and formal analysis of scientific representation, namely ambiguity and convention. The relationship between ambiguity and convention will be examined in detail in the chapters that follow. For now, however, it suffices to make the basic point that ambiguity exists where formal analysis is underdetermined. Furthermore, in order for the social dimension of representation to function, this ambiguity must be resolved by a choice, often known as a convention. Thus, in the conventional resolution of formal ambiguity, one has identified a fundamental way in which the social and formal dimensions of representation interact.

In order to illustrate this further, consider again the information relation. This relation exists between an actor and an audience when there is a faithful transfer of information between them. As it was introduced above, the information relation is fundamentally intended to hold between two individuals. The degenerate case of representing the world to oneself was also briefly considered and rejected on the basis of Wittgenstein's argument for the impossibility of private languages. However, the intuitive sense that it is possible to represent the world to oneself still may have its appeal to some. There is, thankfully, a more direct response that also illustrates the role of convention and ambiguity in reference to the information relation. If the social mediation of scientific knowledge is to involve only a single person, then there is no reason to choose which structure, of a number of similar structures, to designate as the model and which the original, or alternatively which structure to choose as the token of a type. Certainly, the would-be solitary representer may resolve these ambiguities by choice, but it is not clear to what end he or she should do so unless this act is a rehearsal for an attempt to transfer information to others. This suggests the conclusion that if scientific representation involves social mediation of information, then formal ambiguities must be largely resolved in order to facilitate the transfer of information.

Finally, the formal and social dimensions of scientific representation also interact as a result of the so-called "theory-ladenness" of experiment. In the case that a model M represents O, it was suggested previously that the existence of a homomorphic mapping from O to M is necessary but not sufficient for scientific representation. There is still, even here in the realm of a purely formal relation, a sense in which the social dimension of representation impinges. Even though these structures are primarily elements of the formal chain of media, they are also elements of the social chain of media, which includes scientists and apparatus. It is now widely accepted among philosophers of science that data, and hence the apparatus that produces this data, are theory-laden. Thus to the extent that this is the case, the idealized model, *O,* and the mathematical model, *M,* which constitute the formal chain of media, are also encoded into the very apparatus that forms part of the social chain of media used to transfer information across a social network.

In conclusion it is helpful to summarize these three central relations of scientific representation in a single schematic—see (8). This scheme further develops that presented previously in schematic (5). From this perspective, the two dimensions of representation, formal and social, may be related to one another usefully, via three interacting relations; the information relation, the representational relation, and isomorphism.

[. . . → Audience/Actor → Audience/Actor → . . .]

World *(W)* Mind (8)

$[O \leftrightarrows M]$

Here the bold arrows indicate the information relation and the direction of its flow along the social chain of media. The relations between elements of the formal chain of media include structural similarity, especially isomorphism, and the representational relation. The latter is indicated with a single-ended arrow to convey the direction of representation. The former is indicated with a double-ended arrow to display the directional ambiguity inherent with this formal relation.

Scientific Representation and the Arts

The combined account of scientific representation summarized above in schematic (8) is an idealization of a highly complex activity. Specifically, it will be useful as a starting point for discussing objectivity and invariance. In general, idealizations like this have the merit of allowing one to focus on a relatively small number of factors—in this case, three key relations. However, the merit of greater simplicity is at the same time a weakness, since certain aspects must be removed from consideration. This kind of loss is always a hazard when one uses an idealization to understand as complex and subtle an activity as scientific representation.

One possibility which remains, however, is to augment the idealized account presented above with an analogy that helps confer a more comprehensive understanding of the phenomenon in question. In the case of scientific representation the most obvious source of such an analogy is found in the arts. Indeed, there is a strong tradition of drawing parallels between science and the arts.[18] This has almost always been limited to discussions of the visual arts and visualizable scientific representations. Although such discussions certainly have merit, extending the comparison to the performing arts may provide a better sense of the interaction between the social and formal dimensions of representation argued for above. Although it is not strictly necessary for the discussion in the following chapters, pursuing this analogy between science and art, including the performing arts, may be one good way to supplement the idealized account presented above.

The analogy between artistic and scientific representation is a very natural one, especially with respect to the visual or graphic arts. Although not all visual art is representational, the most obvious parallel between science and the arts is that both activities are in large part oriented toward constructing representations of reality. In addition, both science and art rely on highly specialized skills, practices, and instrumentation; one may even speak of the abilities of an especially gifted individual as virtuosic in science as in the arts. One may think of the activities of a physicist representing space and time with the spacetime manifold of relativity theory as doing something analogous to the efforts of an impressionist painter, who uses quite a different set of structures and techniques to represent the world.

This comparison is particularly appropriate when one focuses on examples like this from the visual arts and areas of science that seem to admit accessible mental pictures of the physical models being proposed. In both of these cases, visual artistic and visualizable scientific representation, representation appears to depend entirely on images. These are usually taken to have representational significance regardless of the process of their production and presentation. The painted image of a starry sky, for instance, is simply the representation. This exemplifies a static, or synchronic, view of representation. Thus the study of artistic form in a painting is analogous to the study of the mathematical structure of a scientific model.

As one seeks to extend the analogy to the performing arts, however, certain advantages become apparent. A performance—a theatrical drama, for example—may represent a scenario but often relies on a temporal structure, a plot that unfolds only over time. In contrast to the situation with visual art, representation in performance art is dynamic or diachronic, taking form within the context of time as well as space. It is in this sense that representation in physics is also analogous to performance art, especially in the way that its social and formal dimensions interact. Indeed, the use of the terms "actor" and "audience" above invites just such an analogy.

A performance has a few definitive characteristics that are relevant here. First of all, and most obviously, one performs actions. That is to say that performance involves, and cannot exist without, the action of performers. Second, performance involves actors on a stage: for science this is physical reality, including its temporal and spatial dimensions.[19] Third, to call an activity a performance is to assume the existence of an audience; a performed action differs from a rehearsed one according to the presence or absence of an audience. These criteria—actor, stage, and audience—constitute key characteristics of performance, which apply just as well to the activities of actors and musicians as to the social dimension of scientific practice.

It has already been suggested that the social chain of media relevant to scientific representation includes both scientists and apparatus. The relationship between an actor and an audience, found in performances, is analogous to the way in which scientific knowledge is socially mediated. In the arts, the nature of such a relationship between an actor and an audience might take a number of forms, including empathy, trust, fear, or some other mode of human relatedness. In science, the relationship between representer and audience may involve some or all of these modes on one level or another. Fundamentally, however, representational relatedness in the sciences depends on the transfer of information. This information relation, denoted by arrows in schematic (8), is just that which results in a transfer of information between an actor and an individual member of the scientific audience. As previously discussed, this information is often encoded in terms of the formal relations between structures, among them the representational relation and structural similarity, especially isomorphism.

Another parallel between artistic and scientific practice is that neither all art, nor all physics, is best understood as representational. Considering the example of music, it becomes difficult to say exactly what one means by a representation. Both a chord made up of simultaneously played musical notes and a melodic line clearly have nontrivial structure, but it is not obvious what one might say they represent as compared to a painting of a landscape. In fact, one does not have to look as far as music to notice that some artistic images are apparently not representational at all. Much of contemporary visual art can be interpreted, perhaps like a musical chord, in terms of its structure, hence structuralism as a mode of understanding art. Structuralism as a way of conceptualizing reality is of course not limited to the arts. It appears in the philosophy of science, for example, in terms of structural realism, which in some forms at least seeks to cast physical theory entirely in terms of set theoretic structural relations.[20]

In addition, there are areas in physics in which representational purposes are dominated by other concerns. Two such concerns are predictive power and mathematical elegance. An example of the former may be found in quantum theory. The correct physical interpretation of the theoretical constructs of quantum theory is a matter of extensive debate, ranging from seeing the state vector as representing a real state of some quantum object to treating it as a mere statistical device. Whatever one's view on interpretation of quantum theory, however, it remains a powerful predictive formalism, and its usefulness in prediction has not obviously been diminished by a lack of consensus about how and if it truly represents reality. Furthermore, examples of theories appreciated for their elegance or simplicity abound in the history of science. Most famous among these is perhaps the heliocentrism of the Copernican solar system, which was preferable (at least in part) because of its perceived simplicity relative to Ptolemaic geocentrism.

Not surprisingly, there are also significant differences between representation in science and the arts worth noting. One of these involves the traditional question of whether representations reflect the intentions of the artist, author, or composer. In the arts intentions are often thought to be an important component of a work of art. One can interpret an artistic representation in ways quite unintended by the artist, but scientists' motivation for creating a model are usually more or less clear.[21] The intentions of a scientist in constructing a model are therefore much less open to interpretation. Moreover, they are irrelevant to the analysis above, which explicitly depends on the socially mediated transfer of information.

Having noted this potential disanalogy between representation in the arts and in science, one may still benefit from some of the positive analogies, as a way of complementing the idealized account given above. In particular, it is hoped that a flavor of the true complexity involved in the interaction between formal and social dimensions of scientific representation may be gained through the comparison to artistic performance. Although one can easily think of scientific models as analogous to graphic art, leading to a discussion of form and the relationship between resemblance and representation discussed above, this only captures part of the story. In effect, graphic art is primarily analogous to the formal dimension of scientific representation. Performance art, on the other hand, with its actors, audience, and stage, is a closer analog to the social dimension of scientific representation. Finally, if one takes the somewhat controversial view that graphic art is a special case of performance art, then one can bring together the formal and social dimensions into a single shorthand slogan, "representation as performance."

Whether or not one accepts this final suggestion, the analogy between science and the arts, both graphic and performance art, helps convey a number of aspects of scientific representation not present in the idealized account presented above. In particular, it may also help illustrate the intimate relation between social and formal dimensions of representation, which has already been argued for in terms of three key relations. In any case, this analogy certainly does not take away from the combined account of scientific representation that forms the starting point for the following discussion of objectivity and invariance in the physical sciences.

Conclusions

Scientific representation may be analyzed in terms of a social and a formal dimension. Because of the significant ways in which these dimensions interact, especially as evidenced by the resolution of formal ambiguity via convention, it is important to keep both of them in the frame. For this reason, the combined account presented above is an important starting point for analyzing the claim that objectivity in physical science is related to the invariance features of its models. This claim implicates both the social and formal dimensions of scientific representation and all three of the key relations introduced above. The social dimension is evident from the fact that objectivity is usually understood as relating in some way to the agreement of observers. Formal analysis, on the other hand, is called for in evaluating the invariant features of a given model. The fact that both dimensions of scientific representation are relevant here will ultimately help demonstrate that invariance has as much to do with convention as with objectivity in the physical sciences. This will in turn provide support for the view that the tension between objectivity and conventionality is often only apparent.

This combined account of scientific representation provides a backdrop for the primary focus of this book, understanding the philosophical significance of symmetry in modern physics, and especially evaluating the claim that invariance is linked to objectivity. The following chapter will take the next step toward this end by taking a closer look at the relationship between mathematical models and symmetry.

2

Models, Symmetry, and Convention

This book attempts to account for the philosophical and methodological significance of symmetry principles in physical science. This task is complicated by the fact that the notion of symmetry implicates several closely related concepts, including invariance, objectivity, representation, and convention. Although the literature on the philosophical significance of symmetry/invariance is growing, there remains no clear consensus on how these key concepts relate to one another. This has led to considerable confusion.

In hopes of clearing the ground for a discussion of the true significance of symmetry in scientific representation, this chapter seeks to relate the notions of representation, symmetry, invariance, and convention in a way that does justice to the practice of the working physicist. Having established these relationships, one will be in a position to address the purported connection between objectivity and symmetry in physics (discussed in Chapter 3).

Models and Symmetry

Physicists attempt to represent reality. As previously discussed, one way to conceptualize this project is in terms of structures and the relationships between them. The structures of particular interest are models. When a physicist sets out to represent some particular physical system, *W*, in the world, its scientific representation depends on two more structures, an idealized conceptual model, *O*, and a mathematical model, *M*.

Turning to the central concept of symmetry, its relationship with representation is often discussed. In its simplest form, however, this relationship may be captured in the observation that structures used in scientific representation, like *W*, *O*, and *M*, often have nontrivial symmetries. Before discussing the significance of these symmetries for scientific representation, a word about symmetry in general is in order.

Symmetry has long seemed a fundamental feature of human experience. The term has been used to refer to such issues as proportion and form in art and architecture, which have interested philosophers from antiquity to the present. This general sense of symmetry is relevant to an understanding of the sciences, as evidenced by the significant role of elegance and form within physics. In addition, the solutions of numerous significant problems within the physical sciences have rested on symmetry arguments of one form or another. Among these arguments are those that make use of the term "symmetry" in a more specific sense. Within the history of modern physics and mathematics, the notion of symmetry has been formalized precisely within group theory. In this sense, the word "symmetry" takes on the meaning of invariance under a group of structure-preserving transformations. A symmetry, or automorphism, of *W*, *O*, or *M* is a transformation that preserves all of its structural relations.

Formally speaking, a transformation acts on a structure and maps each element of that structure into the elements of another structure. This may take place in a number of ways (including injections, bijections, or surjections); in one of its simplest forms, "point transformations," Felix Klein explains that these maps involve a one-to-one correspondence between elements that is effectively "a generalization of the simple notion of a function."[1] A transformation may be encoded by specifying which element of one structure is to be associated with which element of the other, and the direction in which change (from the first structure to the second) is meant to take place. This may be indicated schematically as an arrow that has both endpoints and a direction.[2]

Among possible transformations, *symmetry* transformations, or automorphisms (or often simply "symmetries"), have the special feature that the elements of the first structure map back into elements of that same structure in such a way that the structure has been, in a sense, preserved. This is very similar to the case in which one structure is mapped into another identical structure; this latter transformation is called an isomorphism. Both automorphisms and isomorphisms preserve the relationships between elements of a given structure, its shape. However, the automorphism, or symmetry transformation, does this in such a way that the relational features of a particular structure are in fact unchanged after the operation of the automorphism. The relational features of a structure are those that are based only upon the relations between its elements and not the objective identity of those elements. In terms of its relational features, the result of operating on a structure by any automorphic map is the same as if it were left undisturbed (or operated on by the identity). It is in this sense that one often hears it stated that symmetries preserve structure. It is important to point out that the kind of structure so preserved is necessarily abstract in the sense introduced in the previous chapter. In fact, it is the abstract structure that is characterized by the relational features of a particular concrete structure.

Consider a simple example of the symmetries of a model. Suppose that one models a crystalline solid with a cubic unit cell, in which the unit cell is considered to be the three-dimensional motif or repeating pattern in the crystal's internal structure. One may now rotate this unit cell by 90 degrees around an axis through its center and perpendicular to one of its faces. After having performed this operation, the cubic structure will appear unchanged. The angles between faces, 90 degrees, and the number of faces, six, remain invariant. These invariant quantities associated with the unit cell, which may be used to characterize the structure of the crystal, are often manifested in the macroscopic morphological symmetries of the crystal.

In the terms introduced in the previous section, the crystal is the physical system in the world, *W*, and the cubic unit cell is a model, *M*, of an idealized model, *O*, of *W*'s internal structure. That *M* here is invariant under certain symmetry transformations is at once a simple geometrical fact and a significant feature of the scientific representation of this physical system. This captures the simplest sense in which models relate to the concept of symmetry in scientific representation, namely that a given model will respect certain symmetries.

There is another slightly more subtle sense in which symmetry transformations relate to scientific models. Sets of automorphisms may also be used as concrete structures in themselves, to represent specific abstract mathematical structures, the groups of group theory. Both of these uses are relevant to the current discussion, as both are aspects of representational practice in modern physical science. However, the latter use of sets of automorphisms to represent abstract groups is necessarily a narrower endeavor than the use of scientific models to represent the various features of physical reality. Nevertheless, some of the most striking applications of symmetry principles in theoretical physics have stemmed from the assertion that the symmetries of a given model represent some abstract group or another.

Up to now, discussion has focused on individual automorphisms, but in practice they are most often considered in sets. Such a set of automorphisms may be used to represent, as a token of a type, the (nonalgebraic) abstract structure of a specified group. There is a great deal that can be said about the specific representations of given abstract groups in this sense.[3] Without going into too much detail, one may note that in this situation it is the sets of transformations themselves that constitute the structures that are representations of a given group.

Here a group is the technical designation referring to any set that holds the basic defining relations of group theory with respect to one another. These include the requirement that, in order to be a group, a set must be closed under multiplication, associative, have an identity element, and include the inverse of all of its members. Significantly, the result of multiplying a series of elements is not necessarily equivalent to another series of the same operations performed in a different order (in other words the members of the group need not commute). Finally, a given group may be specified by a multiplication table that designates the result of multiplying any two of its members. This abstract structure of a given group is an example of a nonalgebraic structure.

Such an abstract group may be represented (in the token/type sense) by the set of automorphisms of a structure.[4] A set of linear operators in a vector space is said to be a representation of a group if its members satisfy the conditions above. The structure of the group, as suggested above, is characterized by a multiplication table. Recalling that any group must be closed under multiplication, this table specifies what the result of any combination of transformations will be.

Returning to the example of a crystalline solid with a cubic unit cell, if one automorphism, a 90-degree rotation, is followed (or multiplied) by another 90-degree rotation, the resulting 180-degree rotation is also an automorphism of its symmetry group. If this is in turn followed by another 90-degree rotation, the resulting 270-degree rotation is yet another automorphism of the crystal. In this way a whole set of rotational symmetries may be generated. This set will have the feature that any transformation followed by another results in a transformation within the set, that the set contains the inverse of every member, that it contains the identity and is associative; the set of structure-preserving rotations is a concrete representation of an abstract group. One example of how the set of automorphisms of a physical system has been used to represent abstract groups is within the practice of theoretical particle physics. Perhaps the most celebrated historical instance within this tradition is that of the "eightfold way," whereby the solution to the eight baryon problem has symmetries that are a representation of the "SU(3)" group.

Symmetry and Invariance

To take this discussion further, one must come to terms with the relationship between symmetry and invariance. Invariance in the context of physics is inextricably linked to the notion of symmetry. Indeed, the terms are often used interchangeably. In reality, the distinct notions of symmetry and of invariance are conceptual duals that rely upon one another. As previously suggested, the symmetries, or automorphisms, of a structure like some model, *O*, are those transformations that preserve its structure. Meanwhile, the aspects of structure, *O*, which are so preserved are called its invariants. This dual relationship between symmetry and invariance means that speaking of one automatically implicates the other.

Since the automorphisms of a given structure form a group (or, more precisely, the concrete representation of an abstract group), one may summarize the relationship between symmetry and invariance by noting that a structure *O* is invariant under its symmetries; that is to say under its automorphism group. This leaves open the question of whether one of either the symmetries or the invariants of *O* is somehow more fundamental than the other. Do symmetries depend on or perhaps explain invariance in some way, or is it the reverse?

A considered response suggests that neither can be the case, since logically speaking these dual terms depend equally on one another. Claiming that there is a duality between symmetry and invariance implies that each is defined in terms of the other and that neither concept may be reduced to the other. The paradigm example of duality in modern physics is the wave-particle duality of quantum theory, according to which one may assert (the apparent contradiction) that the same physical system acts like both a wave and a particle. This is a duality because the behavior of the system may not be reduced to waves on the one hand or particles on the other. A similar notion of duality occurs in geometry, in which dual entities are exemplified by, for instance, a point and a line; in this case infinite localization and infinite extension "always enter symmetrically" in terms of one another.[5]

Perhaps the most fundamental duality evoked in this discussion is that between change and permanence. This duality has a long history in philosophy and in its simplest form may be expressed in the claim that permanent entities are precisely those that endure through change; conversely change is only evident when some aspects of these entities fail to endure. This duality finds a formal expression in terms of the dual concepts of symmetry and invariance.

Practically speaking, however, there is a strong tendency to treat one or the other as fundamental. Felix Klein, whose 'Erlangen program' is usually credited with introducing the conceptual apparatus of group theory to the study of invariance, classified mathematical structures and geometries in terms of their symmetries. Within the physical sciences, however, Poincaré and Lorentz, in their study of electromagnetism, assumed the representation of electromagnetic waves via Maxwell's equations as invariant, and derived symmetries, the Lorentz transformations, which preserve Maxwell's equations.

On the other hand, in physics as with Klein in mathematics, one may start with a set of symmetries and derive invariants. Einstein's presentation of special relativity did just that, moving from symmetries, including the Lorentz transformations, to the invariant spacetime interval "proper time." In doing so, Einstein successfully extended symmetries of electromagnetism to make predictions about mechanics. In fact, the elegance and power of this approach accounts for a great deal of the interest physicists have had in applying a group-theoretical understanding of symmetry and invariance to their theorizing.

Regardless of which approach one takes, starting with symmetries and deriving invariants or the reverse, there is in any given circumstance a distinctive inverse relationship between the two: the greater the number of symmetries, then the smaller the number of invariants and vice versa. Intuitively, a structure with more invariant features will have less symmetry, and a highly symmetric structure will have fewer invariant features. Putting it in group-theoretical terms, for some structure, *O*, the larger its group of automorphisms, the smaller its set of invariants.

To illustrate this point, one may imagine how a group, a set of symmetry transformations, may be built up—the number of members of the group increasing as new symmetries are added. With each addition, each new group of symmetries leaves a different set of structures invariant. These sets of invariant structures are sometimes called "orbits." In general, the invariant features of these structures are inversely proportional to the size of the group of symmetries under which they are invariant. This appears, for instance, in one of the classificatory schemes used for crystal structures: Triclinic, monoclinic, and cubic unit cells are members of different orbits (see Figure 2.1).

The characteristic vectors specific to each one of these crystal systems have different numbers of invariant angles and lengths, inversely proportional to size of the groups of transformations under which they are invariant. Thus a cubic unit cell is characterized by three orthogonal vectors of the same length; its symmetries include fourfold rotational symmetry about three orthogonal axes passing through its center. A triclinic unit cell, by contrast, is characterized by three vectors of no fixed ratio of lengths or angles with one another. It displays only onefold rotational symmetry. Thus the structure with more independent invariant features displays less symmetry.

There is no doubt that symmetry and invariance play a central role in the development of modern physics. However, a lack of clarity as to the relationship between these notions has led to some confusion, especially in the philosophical literature that seeks to account for their significance. It is important to recall that these are truly dual notions and that, formally speaking at least, neither invariance nor symmetry is more fundamental than the other. In addition, it is helpful to keep in mind the significant inverse relationship between the two.

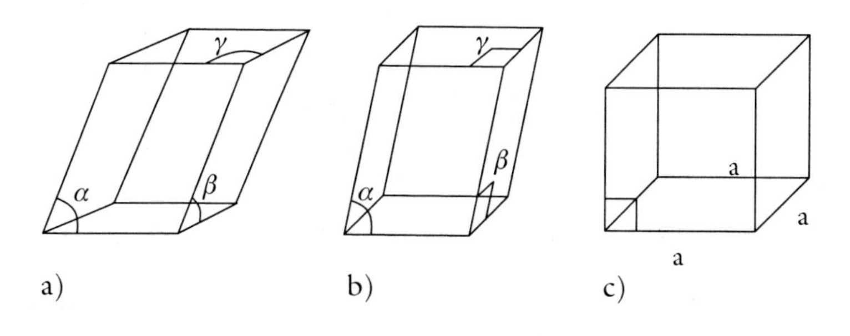

Figure 2.1. a) triclinic unit cell, b) monoclinic unit cell, c) cubic unit cell.

Symmetry and Ambiguity

One may begin to account for the methodological significance of symmetry in scientific representation by noting that symmetry introduces structural ambiguity. This ambiguity may be characterized differently depending on whether one adopts so-called "active" or "passive" interpretations of the relevant symmetry transformations. In any case, however, the ambiguity introduced by symmetries must be settled by conventional choice. Thus, the first sense in which a scientific model's symmetries are important has to do with the scope they allow for conventional choices.

As previously suggested, a symmetry, or automorphism, of structure *O* may be interpreted actively or passively. These interpretations are formally equivalent in the sense that in each case the relevant transformations may be expressed in terms of the other. Understood actively, an automorphism involves a map of one structure onto itself within a given mathematical representation. By contrast, a passive interpretation takes the automorphism to be a map between two mathematical representations of the same structure such that all of its features are preserved.

These interpretations are, in reality, formally equivalent. In each case the relevant transformations may be expressed in terms of the other. Another way of expressing the difference is to note that a passive symmetry involves a map between two ways of coordinatizing a single structure whereas an active symmetry involves a mapping between elements within a single coordinate system.[6] Intuitively, the first is passive because it involves a single structure viewed from different perspectives and the second active because the transformation directly moves the elements of the structure back into itself. Thus, Klein explains, under a passive interpretation, a transformation indicates "a change in the system of coordinates," whereas an active interpretation the transformation "holds the coordinate system fast and changes space."[7]

It is important to note that, formally at least, there is an equivalence between these two interpretations. Transformations between two isomorphic structures and automorphisms of a single structure may be factorized in terms of one another. Redhead has previously summarized the basic relationships between isomorphisms, automorphisms, and structure by relating a physical structure, *P*, to a mathematical structure in terms of the following lemma: "Any symmetry of *P* can always be factorised in the form of $y^{-1} \circ x$ [*x* followed by the inverse of *y*] for suitable choice of the maps *x* and *y*."[8] Here *x* and *y* are mappings from *P* to *M*. They have the property that they map the structure of P onto M exactly, making them isomorphisms. Thus, denote the symmetry of P by the map $\phi : P \rightarrow P$. Now form the map $y \circ \phi : P \rightarrow M$, where y is any given isomorphism of P onto M. This is clearly also an isomorphism—call it x; so, $y \circ \phi = x$, or $\phi = y^{-1} \circ x$; QED. See Figure 2.2.

This lemma indicates that the combination of one of these transformations followed by the inverse of the other will result in a map of P back onto itself; such a structure-preserving map of one structure into itself is an automorphism, here understood as an active symmetry transformation. In just the same way, symmetries of M, coordinate transformations, or passive symmetries take the form $y \circ x^{-1}$.[9] So if one denotes by Sym(P) the group of all symmetries of P, and similarly for Sym(M), then Sym(P) and Sym(M) are abstractly one and the same group represented in distinct ways by the members of Sym(P) and Sym(M). Since the mappings between P and M are isomorphisms, it is no surprise that the automorphisms of each structure, active and passive symmetries respectively, may each be factorized in terms of x and y.

Their formal equivalence notwithstanding, the passive interpretation of symmetry is perhaps more helpful in illustrating the type of representational ambiguity that arises from the symmetry of O. The connection here between symmetry and representational ambiguity leads one directly to

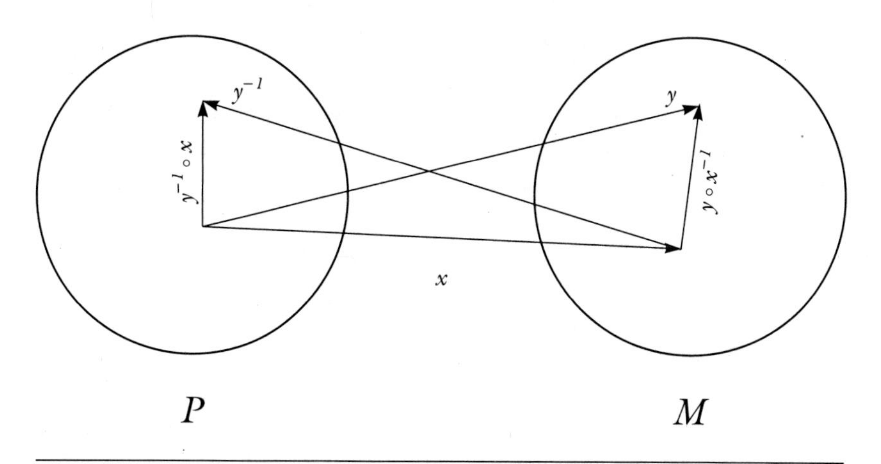

Figure 2.2. $x : P \rightarrow M$ and $y : P \rightarrow M$ are distinct isomorphisms between a physical structure P and a mathematical structure M. $y \circ x^{-1} : M \rightarrow M$ is a coordinate transformation, while $y^{-1} \circ x : P \rightarrow P$ is a symmetry of P.

the role of conventional choice in scientific representation. If a symmetry transformation is understood as mapping elements of one mathematical representation to another, then one may just as well choose either representation. In this way symmetries introduce ambiguity that must be resolved by conventional choice in the way scientific models are used.

The most obvious example of this within twentieth-century physics is that of the ambiguity existing between inertial reference frames within special relativity. In the terms previously introduced, one may think of special relativity as proposing an idealized model, *O,* of actual events in space and time. This is represented by a mathematical model, *M,* of these events and their spatiotemporal relation to one another in the form of a manifold. It is central to special relativity that certain relations on this manifold are invariant under the Poincaré group (also known as the inhomogeneous Lorentz group), which includes spatial and temporal translations, rotations, and Lorentz transformations. The notion of an inertial reference frame is a way of coordinatizing the spacetime manifold, *M,* using a set of space and time axes. It is simple to show that each inertial frame may be related to an infinite number of other reference frames via the Lorentz transformations. (This will be discussed in detail in Chapter 4). Since these coordinate transformations are automorphisms of the structure of the manifold, then one may equally well choose any inertial frame to represent the structure of spacetime. Resolving this ambiguity involves a choice: One must choose a frame of reference, but it matters not which one.

To put this in general terms, consider a model, *M,* used to represent an original, *O.* If *M* and *O* are isomorphic, then their respective sets of automorphisms are representations of the same abstract group. In this case the automorphisms of *O* are normally called its "symmetries," whereas the automorphisms of the model *M,* also formally symmetries, are usually referred to as "coordinate transformations."[10] In other words, if *M* is used to represent *O,* then it may be used equally well in different orientations related to one another by its automorphisms.

How might this look according to the combined account of scientific representation proposed in the previous chapter? Consider an idealized example of representation in which the formal chain of media includes two isomorphic circular structures, *O* and *M.* One may think of them as intended to model a planetary orbit. Thus the physical system, *W,* is the actual orbit, the idealized model *O* is a circle (an algebraic abstract structure, in terminology introduced in Chapter 1), and *M* is a particular concrete

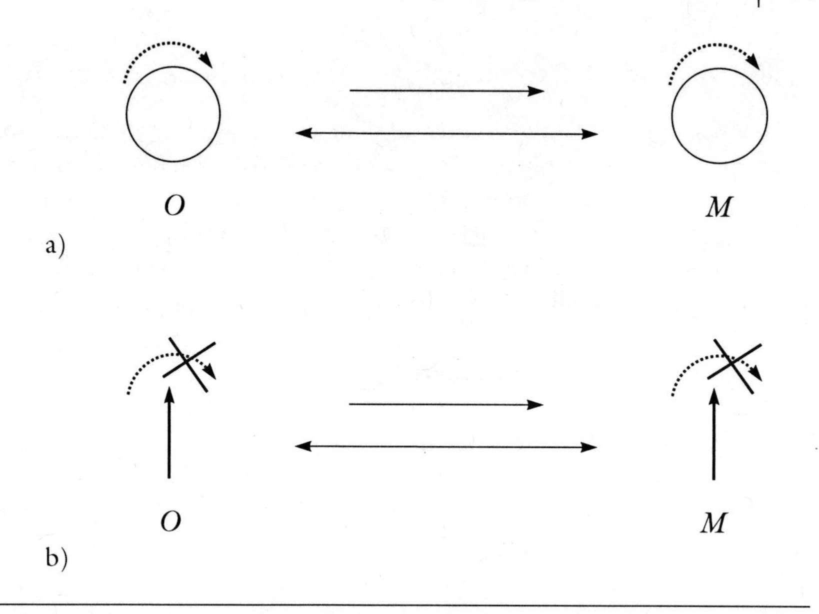

Figure 2.3. (a) Idealized model *O* and mathematical model *M* are here related to each other by isomorphism (depicted as a double arrow) and the representational relation (depicted as a single arrow). In this case all rotations (depicted as dashed arrows) are automorphisms. (b) Idealized model *O* and mathematical model *M* are here related to each other by isomorphism (depicted as a double arrow) and the representational relation (depicted as a single arrow). In this case rotations (depicted as dashed arrows) are not generally automorphisms; only the 360-degree rotation is.

model (see Figure 2.3). The solid double arrow indicates that the structures are isomorphic. The single arrow, the representational relation, indicates that *M* represents *O*. Symmetry transformations (automorphisms), depicted as dotted arrows, lead to the need for conventional choices over how to orient each circle within the chain of media. Simply put, the circles may be used in any orientation.

One may also note that there is a degenerate version of this example in which there are no symmetry transformations other than the trivial case of the identity. Such an example may be constructed by replacing the circular media above with vertical arrows (see Fig. 2.3). Since there are no rotational symmetries, there is no conventional choice over orientation, but the representational relation is still evident.

Ambiguity and Convention

Ambiguity in scientific representation demands an appeal to convention. Convention and especially conventionalism are terms with numerous associations and a complex history. The purpose here is not primarily to clarify this history. One may, nevertheless, note some of the various views on how conventionalism impacts the sciences. Having just introduced the notion of group-theoretical invariance, it is important to note the connection between conventionality and invariance. Invariant structures introduce the ambiguity that calls for conventional choices in scientific representation.

In addition to philosophical approaches to conventionality in science, one may simply note some different types of conventional choice that appear in the sciences. Most generally, if one allows that social interaction in human culture is strongly dependent on conventions, then this is also true of the subculture of practicing scientists. Following from this, it is difficult to dispute the claim that the physical sciences depend on choices that are made by convention: what language to use, how to define specialized terminology, what units of measurement to use and how to define them—an almost innumerable list of such everyday decisions. Because scientific practice is so dependent on these decisions, it is clear that the way scientists represent physical reality is, in this sense, conventional.

Over and above these obvious conventions, conventional choices are necessary in resolving the representational ambiguities. One may divide these choices into those that are conventional in absolute, trivial, or relational senses (see section below, Different Kinds of Convention). Especially significant is absolute conventionalism, which refers to cases in which there is no fact-of-the-matter in preferring one representational convention over another. This will be discussed in detail in Chapter 4 with reference to the debate over the conventionality of simultaneity in special relativity.

Historical Views on Convention

The term "conventionalism" is most often associated with French physicist and mathematician Henri Poincaré (1854–1912). Poincaré's understanding of the use of conventions in geometry and the physical sciences has been an issue of debate from his own day up to the present.[11] The fact that it was not Poincaré himself who introduced the term "conventionalism" suggests that one might maintain a distinction between his intentions and those of others who have found his proposals suggestive for their own research agendas.[12] Nevertheless, Poincaré's legacy, regardless of original intent, has determined some major features of subsequent philosophical discussion about the use of conventions in physics. Poincaré's conventionalism is, furthermore, couched in terms of mathematical structures in a way that is easily applicable to a discussion of models in scientific representation.

With this in mind, one may note a few features of this tradition. Poincaré's view of scientific knowledge may be summarized as follows: "Hard facts . . . are completely independent of the scientist's will. In order to report such facts . . . scientists must agree on certain conventions."[13] Such conventional choices might typically involve the way to construct a spatio-temporal system of geometric coordinates. Most important, according to Poincaré, one may not choose a particular geometry (Euclidean or non-Euclidean for example) over another as a matter of fact. There is, in essence, nothing about the world that requires it to be represented using familiar Euclidean geometric structures. This is a surprising claim: Euclidean geometry is the geometry of our experience. Immanuel Kant gave expression to this intuition in looking to Euclidean geometry as a paradigm of what he famously called the "synthetic a priori," where a priori indicates independence from sense perception. This so-called "apriorist" view of Euclidean geometry as the geometry of human experience met with particular criticism upon the discovery in the nineteenth century of non-Euclidean geometries. Poincaré's conventionalism, coming as it did in the decades following this development in mathematics, asserted the usefulness of any geometry in representing reality. Whereas Poincaré seems to have been primarily interested in conventions having to do with the geometry of space, the question of what geometrical structures to use carries over into other contexts of scientific representation.

Although it is not often discussed in the context of his geometric conventionalism, Poincaré's philosophy of geometry was closely tied to notions of symmetry and group-theoretical invariance.[14] As will be discussed in detail, a number of physicists and philosophers have expressed their suspicion that objectivity and symmetry are related to one another. Significantly, there is no comparable trend that celebrates the link between invariance and conventionality. For this reason, it is particularly important to note that Poincaré spoke in terms of just such a relationship. In comparing, for example, Euclidean and non-Euclidean geometries, Poincaré couched his discussion in terms of the mathematical groups associated with each of them; the Euclidean group includes among its members spatial translations and rotations. In considering this case, Poincaré concludes that "geometry is nothing but the study of a group," and that "the truth of the geometry of Euclid is not incompatible with the truth of the geometry of Lobachevsky, for the existence of a group is not incompatible with that of another group."[15] Thus, Poincaré compares the choice of a group to what is in effect a choice of coordinate systems, a conventional choice that resolves coordinative ambiguity. Although here one sees the choice of a group as essentially equivalent to the choice of a geometry, Poincaré's conventionalism has more often been discussed with emphasis on the latter. Moreover, this is an excellent example of how one may view the choice of a relevant mathematical group as a geometric convention.

For Poincaré, the conventional choice of a group, and thus a geometry, should be made on the basis of practical considerations. To cite one of his examples, Poincaré attempts to account for why humans have so often chosen the Euclidean group and Euclidean geometry to represent the spatial aspects of our experience. He first of all remarks that "there exist in nature some remarkable bodies which are called *solids,* and experience tells us that the different possible movements of these bodies are related to one another much in the same way as the different operations of the chosen group."[16] As previously suggested, the set of automorphisms of a physical system may be used to represent an abstract group. This approach is perfectly consistent with Poincaré's approach to geometry. In trying to explain how humans might have first discovered geometries, and why they might have tended to think in terms of Euclidean geometry, he advances a sort of argument by natural selection according to which this geometry is the most effective in defending human beings against physical attack. As the thrusts and parries involved in combat are the relevant motions of this geometry, then it makes the most evolutionary sense to adopt a three-dimensional geometry that accommodates "unvarying solids."[17] These solids are, of course, invariant structures under the Euclidean group, mentioned above. It is worth noting that this proposal that humans evolved into Euclidean geometers provides an interesting alternate explanation for the geometric intuitions that Kantian apriorism maintains are innate. The main point here, however, is that conventional choice of geometry (or of a group) may be based on the sorts of symmetry transformations and associated invariants that make practical sense for the task at hand, be it self-defense or representing reality to a given audience.

Hans Reichenbach, writing in the 1920s, picked up Poincaré's theme of conventionalism in his book *The Philosophy of Space and Time*. Reichenbach and his successors, especially Adolf Grünbaum, were responsible for perhaps the most discussed conventionalist claim in twentieth-century physics, that of the conventionality of simultaneity. According to this claim, to be discussed in detail in Chapter 4, the simultaneity relation between spacetime points in Einstein's special theory of relativity carries with it an irreducibly conventional element.

While Reichenbach accepted Poincaré's geometrical conventionalism at one level, he felt that he had overlooked "the possibility of making objective statements about real space in spite of the relativity of geometry."[18] Given Poincaré's insistence on allowing for "hard facts" as well as conventional choice, it is initially difficult to see how Reichenbach might come to this conclusion. Picking up the conventionalist thread, Reichenbach begins with a definition for the "principle of the relativity of geometry" (which he credited in part to Poincaré): "Given a geometry G' to which the measuring instruments conform, we can imagine a universal force F which affects the instruments in such a way that the actual geometry is an arbitrary geometry G, while the observed deviation from G is due to universal deformation of the measuring instruments."[19] This is, in fact, the sort of expression of conventionalism that has often been advanced as characteristic of Poincaré.

To this Reichenbach adds an important notion of his own, the coordinative definition, by which abstract mathematical structures are intended to be related to physical objects in the world. It is likely that Poincaré's omission of such a step, choosing to restrict his discussion to mathematical structures, is the basis of Reichenbach's concern that his geometrical conventionalism had denied the possibility of making objective statements. Thus Reichenbach seems to have been using objectivity in the sense of ontological subject independence.[20]

Reflecting this distinctive approach, Reichenbach expressed his conventionalist program by asserting, "Properties of reality are discovered only by a combination of the results of measurement with the underlying coordinative definition."[21] In his terminology, a coordinative definition is used to link a "concept" to a physical entity, for example the concept of distance to a physical measuring rod.[22] Coordinative definitions are in essence another form of convention, but ones that "lend an objective meaning to physical measurements."[23] Therefore Reichenbach is able to conclude, "The objective character of the physical statement is thus shifted to a statement about relations."[24] That is to say, it is shifted to those relations specified by coordinative definitions.

Quite apart from the tradition that follows from Poincaré through Reichenbach, another particularly influential approach to conventionality in science has been phrased in terms of so-called "under-determination" of theory by data. Pierre Duhem, a philosopher and physicist who published in the early decades of the twentieth century, is associated with certain instrumentalist ideas regarding scientific theory. Some of these were later taken up again by the philosopher of science W. V. Quine under the name of the Duhem-Quine thesis.[25]

Many philosophers of science have come to conclude that scientific theory is in general radically underdetermined by the data, or as Quine put it, "The totality of our so-called knowledge . . . is a man-made fabric which impinges on experience only along the edges."[26] As another way of illustrating the same point, the theory-laden feature of the instrumentation, by means of which data is collected, implies that any specific new datum can be accommodated by any number of changes in the rich network of theoretical structures that link observation to theoretical representation. The choice of where in the network to make the adjustment is simply a matter of convention. Quine suggests that there is a "natural tendency to disturb the total system as little as possible."[27] This approach has been hugely influential in the history and philosophy of science. Here it will suffice to point out that experiences are in Quine's view quite real; they retain a structure that, though not sufficient to determine our complex fabric of conventional representations, is nevertheless objective in an observer-independent sense. If this were the case, then one might choose by convention the way in which reality should be modeled, without undermining the objectivity of our experiences.

Different Kinds of Convention

Having considered some elements of the conventionalist tradition, one may also consider some types of conventional choice that appear in scientific contexts. The suggestion that something involves a conventional choice in its most practical sense indicates that individual scientists must agree among the relevant research communities on various standards. This practical sense of convention is not primarily what proponents like Reichenbach and Poincaré intended. One may not always, at least in principle, need to caucus the scientific community in order to settle on a unit to measure distance. In practice, however, the network of such choices is complex enough that there may be more of a requirement to adopt the existing standards of the community.

As a starting point, some of these choices may be entirely unconstrained by the physical systems under consideration, or they may have constraints that reflect substantive aspects of the physical models being proposed. It is apparent that the choices one makes in the description of physical systems in physics can be considered as conventional in at least three different ways—trivially, absolutely, or relationally.

If a conventional choice is not constrained by the particularities of the physical system being represented, then it may take one of two forms, trivial or absolute. First of all, the choice may be unconstrained because it is a trivial or semantic one, in a sense external to the theory; in this case a theoretical description or argument goes through in exactly the same way but using different labels. An example of this might be the choice of a name used to refer to a specific object or process. The example usually advanced for something like this is that of the translation of terms from one language to another. It seems clear that all physical theory, indeed all linguistic expression, is subject to this sort of trivial conventionalism.

Alternatively, the choice may be unconstrained because the theory itself seems to imply there is no fact-of-the-matter that distinguishes one possible choice from another. In this case, absolute conventionality, the various representations of an original structure differ, but do so in a way that is not constrained by the physical system being studied. The prime example of this in the literature comes from the conventionality of simultaneity, according to which there is no fact-of-the-matter to distinguish a single simultaneity criterion from any other. The substantial difference between trivial and absolute conventionalism comes down to the issue that while both involve conventional choices that are not constrained by the theory, under absolute conventionalism there is, in a Pickwickian sense, significant factual content to this reality. In other words, it is an interesting fact that there is no fact-of-the-matter.

This leaves the third category of conventionalism, relational conventionalism. This is a conventionalism that has to do explicitly with physical systems in the world. A paradigmatic example of this is a length standard, which is set by convention in relation to a standard meter. As already discussed, this is precisely the example that Reichenbach gives to illustrate his concept of a coordinative definition, which on its own captures a basic sense of relational conventional choice.

Of these types of convention, it is absolute conventionality that is of greatest interest here. After all, trivial conventionality is just that, a trivial and relatively easy-to-understand aspect of scientific representation. Relational conventionality, exemplified by Reichenbach's coordinative definitions, is far from trivial, but is also a straightforward feature of scientific practice. Absolute conventionality, with its claim that certain ambiguities may not be resolved by appeal to any fact-of-the-matter, requires a deeper explanation, and among these forms of conventionality seems to be most in tension with the claim that scientific representations are objective. Note that the ambiguity in representing P by the *different* structures M and M' does not in general imply absolute conventionality; this arises only in relation to the ambiguity within a *given* structure. Most significantly, the symmetries of scientific models introduce ambiguities that often must be resolved by just this type of purely conventional choice. That this is the case will be illustrated in detail.

Representation, Symmetry, and Convention

In Chapter 1 a combined account of representation was advocated, one that considered both the social and formal dimensions of scientific representation. The interaction between these dimensions was emphasized, especially as evidenced by the resolution of formal ambiguity via convention. In this chapter, this approach was extended in order to relate the use of scientific models to the notions of symmetry and invariance. It was argued that in general symmetry introduces ambiguity and therefore the need for conventional choice. This led into a more substantial assessment of the conventional element of scientific representation. Once the relationship between representation, symmetry, and convention is established, one is in a position to address the philosophical question of greatest interest here, the purported connection between objectivity and symmetry in physics.

One of the clearest examples of the interaction between social and formal dimensions of scientific representation is found in the ambiguity over the direction of representation, previously discussed. Accordingly, it was pointed out that, given isomorphic structures M and O, it is the social context which supplies the information that M represents O and not the reverse. In addition there is a specific type of ambiguity of special relevance to the discussion of symmetry in scientific representation, the resolution of which also exemplifies the interaction between the social and formal dimensions of scientific representation. This type of ambiguity derives from the symmetries of *M* when *M* is isomorphic to *O*. Thus, the automorphisms of a model *M* actually provide a means of specifying certain types of ambiguity in using *M* to represent *O*. In this way the symmetries of a model are inextricably linked to the scope for conventional choice available to the physical scientist.

In approaching the more specific question of the philosophical significance of symmetry in the sciences, there are several key points to remember in regard to the interplay among representation, symmetry, and convention. First of all, scientific representation may be conceptualized in terms of three structures, a physical system in the world, an idealized model, and a mathematical model; this treatment allows one to bracket the question of how one correlates an idealized model with mind-independent reality and focus on the ways in which mathematical models represent idealized conceptual models. It is in this connection that symmetry principles are applied, and in this connection that they should be philosophically assessed. Secondly, structural similarity is necessary but not sufficient to establish a representational relation between two structures, *M* and *O*. Instead, one must look to the context according to which the physicist uses structures to represent the world. Finally, the representational ambiguities a physicist faces include the ambiguity that arises from the symmetries of the mathematical model. These must often be resolved as a matter of the arbitrary choice of a convention.

3

A New Appraisal of Symmetry

In the preceding chapters, attention has been drawn to several key themes that surround the discussion of the philosophical significance of symmetry in modern science. These themes include invariance, representation, convention, and objectivity. Having now provided an account of the relationships among representation, symmetry, and convention, as well as a brief discussion of the subtle relationship between symmetry and invariance, there remains the issue of objectivity.

Judging from the literature on symmetry, its purported connection to objectivity is perhaps the theme of greatest philosophical interest. A survey of the literature reveals three major senses in which symmetry has been taken to be fundamental to physical science.[1] First of all, one is confronted with a wealth of testimony to the remarkable heuristic power of symmetry in the development of better scientific models in many branches of modern science. Second, one finds that many view symmetry as a means to that great prize of modern physical science: universal theories. Third, one finds a number of philosophically tantalizing references to symmetry as a means to objectivity. Moreover, this third approach seeks to encapsulate and account for the first two. In this sense it also attempts to draw support from them. In addition, proponents of this last view imply that the satisfaction of certain symmetry criteria provide one with a methodological tool in the form of a condition for objective representation. This view, which will here be termed "invariantism," thus represents the most significant extant effort to account for the philosophical significance of symmetry and invariance.

Although a number of philosophers and physicists have been attracted to invariantism, it faces some serious objections, which will be considered here in detail. Before rejecting invariantism as a general program, one must get a better sense for its claims. In particular, it will be shown how it seeks to incorporate and draw support from the heuristic success of symmetry, on the one hand, and the connection between symmetry considerations and unification, on the other.

The rejection of invariantism in its full sense is only the first step in a new appraisal of the philosophical significance of symmetry in modern physical science. The second step will be to suggest that in spite of the failure of invariantism there are certain very specific circumstances under which symmetry criteria may indeed be used as a condition for objectivity. In fact, these cases of what may be termed "perspectival invariantism" are often precisely those cited as concrete examples of invariantism at work. Understood correctly, however, these cases should not be understood as buttressing the general claims of invariantism, but rather as illustrations of the fact that symmetry in representation has as much to do with convention as it does with objectivity.

Thus the new appraisal of the philosophical significance of symmetry in physics offered here involves first of all a characterization and refutation of the extant claims that may be considered under the heading of invariantism. Second, it provides the justification for a limited form of perspectival invariantism, which accounts for a number of important cases in the history of modern physics. Third, and finally, these cases may be used to illustrate the relationship between symmetries and representational ambiguity, thus establishing the more general philosophical point that where symmetry does indicate objectivity, it also implicates convention.

Heuristic Power

The most obvious feature of the literature on symmetry is that it has been used to brilliant effect in the framing of new and powerful models in many different branches of physical science. Famous examples include everything from the standard model of the atom to the molecular structure of DNA to various conservation laws.[2] The heuristic power of symmetry methods in developing new scientific models often depends on a procedure whereby the symmetries of an existing model are generalized and extended to new physical phenomena. Considered in a little more detail, and making use of some of the terminology previously introduced, one may observe that physicists often seek to construct new models by applying the symmetries of one model, O_1 (of some physical system W_1) to features of another O_2 (of some physical system W_2). If O_2 does not respect the symmetries of O_1, then it is replaced by a revised model O_2', which does.

In practice, this revised model is sometimes found to be a better model of W_2 than the original O_2, evaluated according to empirical criteria; that is to say it is taken as a better model of the physical system in the world, W_2. When this happens, the end result appears doubly significant; one has a better idealized model of W_2 (O_2'), and one has in some sense unified the models of W_1 and W_2, in that O_1 and O_2' respect the same group of symmetries. This heuristic "leapfrog" tactic has had a powerful influence on modern physics.

According to its proponents this method may be repeated, leaping from one area of physical science to the next, resulting in a number of models that represent different physical systems. In each leap, a new and empirically fruitful model is produced. Moreover, each of these models is by construction invariant with reference to a continuously revised set of automorphisms, often called the "symmetries of nature."

The history of modern physics is replete with illustrations of just this leapfrog approach. Consider, for example, Einstein's special relativity. In his 1905 paper on the electrodynamics of moving bodies, Einstein famously made the assumption that both the laws of physics and in particular the speed of light are the same in any inertial frame of reference.[3] They are, in other words, invariant under changes in inertial frame. It was known, through the work of H. A. Lorentz and others, that the Lorentz transformations mapped one inertial frame into any other and preserved the model of electromagnetism provided by Maxwell's equations. Thus it may be said that Einstein began by assuming a symmetry, the Lorentz transformation, which left the model of electromagnetic phenomena invariant.

Einstein went on to introduce a new model of space and time and required that his new notion, spacetime, also be invariant under the symmetries of electromagnetic phenomena. In taking this step, intervals of pure space and pure time, not invariant under a group of transformations that include the Lorentz transformations, were no longer considered as objective spatial and temporal measures. Thus, space and time were abandoned as objective features of spacetime and replaced by proper time, which is invariant under Lorentz transformations. Furthermore, extending the requirement for Lorentz invariance to mechanics also exemplifies a heuristic leap. Since the original equations do not exhibit this invariance, doing so introduces new terms with the result that relativistic mass is no longer constant but depends on velocity.

Taking another heuristic leap, quantum mechanical models of particles may be assessed according to invariance under the automorphism group of special relativity. The automorphisms of special relativistic (Minkowski) spacetime form the Poincaré group, which includes rotations, translations, and Lorentz transformations. Once again, however, when one applies these symmetries to quantum mechanical models, certain features of these quantum states fail to be invariant under the Poincaré group—for one, their localizability. Therefore, these features of quantum mechanical models may be altered as well; in practice this has been realized by the development of quantum field theories (to be discussed in detail in Chapter 6).

It is a fact of history that physics has often advanced in this manner. As a result, physics has produced models of much of the physical world which are invariant under a group of transformations that are considered to be symmetries of nature. The empirical success of these models presents a dramatic case in favor of the heuristic leapfrog method. In addition, as previously suggested, the fact that the collection of models so constructed is invariant under the same group of symmetries provides physicists with a strong sense that they are unified.

Universality

Perhaps not surprisingly, another sense in which symmetry has been held to be fundamental, especially within the physics community, is in its connection with the project of constructing increasingly universal or unified theories. For example, the Dutch physicist Vincent Icke suggests that "physicists believe that the underlying symmetry, which forbids whole classes of occurrences [of phenomena] at one stroke, is in a sense more fundamental than the individual occurrences themselves, and is worth discovering."[4] If symmetry forbids classes of events, then it may be said to discipline our representations of reality in a way that moves away from the particular to the universal. Reflecting similar views in his popular book, *Dreams of a Final Theory,* Steven Weinberg claims that modern physics "has evolved a new view of the world, one in which matter has lost its central role. This role has been usurped by principles of symmetry"; furthermore a "final theory will rest on principles of symmetry . . . [that] will unify gravitation with the weak, electromagnetic, and strong forces of the standard model."[5] Weinberg's view of symmetry as a means to unification is in this way essentially similar to Icke's. Both use symmetry considerations to move scientific representation away from the particular. This approach is, at least implicitly, accepted by many within the physics community. To cite one more example, Sunny Auyang uses this very terminology by noting that "Symmetry transformations erase particularities" in favor of universals.[6] The consideration of symmetry in physical theory may thus be seen as a guide to universal statements about reality.

Related to this view is the claim made by Bas van Fraassen that scientists use the terminology of symmetry, transformation and invariance, to capture the universality and necessity traditionally expressed through the concept of laws of nature.[7] In fact, van Fraassen's book *Laws and Symmetry* is one of the first major efforts by a contemporary philosopher of science to come to grips with the role of symmetry in modern science. One of the most clear-cut examples that he cites of a link between symmetry and laws is found in the well-known methodological principle first proposed by Emmy Noether. Referring to this case, among others, Eugene Wigner concludes, "To be touchstones for the laws of nature is probably the most important function of invariance principles."[8]

Van Fraassen is skeptical about the suggestion that there are such necessary and universal laws in science. He concludes that "the conceptual triad of symmetry, transformations and invariance does not explicate or vindicate the old notion of law."[9] From this it might appear that he is rejecting the notion that symmetries are related to universality in representation. This is not the case, as he also concludes that in symmetry one "has indeed found a significant notion of true generality, but not one of necessity."[10] His final position is very much within the tradition maintaining that there is a connection between symmetry and universality. Thus, van Fraassen rejects the claim that symmetry accounts for laws of nature as spurious because laws demand necessity in addition to universality.

Objectivity

Perhaps the most philosophically significant approach to symmetry has been to view it as a guide to objective statements in science. Taking quantum field theory as her example, Auyang has expressed this approach by claiming that "modern physical theories make explicit the meaning of 'objective'" by making statements about quantities that are invariant under relevant groups of symmetry transformations.[11] Similarly, Peter Kosso has recently suggested that objectivity "can be securely linked to symmetry."[12] The two names associated with early versions of this sort of proposal are those of the nineteenth-century mathematician Felix Klein and the early-twentieth-century physicist Hermann Weyl. In fact, Weyl has provided the simplest version of this view in the claim that "objectivity means invariance with respect to the group of automorphisms."[13] A more recent version of this approach has been argued in detail by Robert Nozick.

From this perspective, the observation that symmetry leads to universality appears to dovetail nicely with the leapfrog account of the heuristic use of symmetry. Accordingly, these factors are often taken implicitly or explicitly as evidence in support of the broader philosophical claim that invariance under a group of automorphisms is a sufficient condition for objectivity in the sciences and beyond. On this account, it is the objectivity of a representation that underpins both the heuristic success of a given model as well as its universality. Versions of this view may be grouped under the heading "invariantism," because of the central role invariance is intended to play both ontologically and epistemologically. Simplifying Weyl's claim even further to the slogan "objectivity means invariance," it should be obvious that much turns here on how one understands objectivity. Thus, before looking at invariantism in detail it is worthwhile to address the question of what is meant by objectivity.

Opposites of Subjectivity

Objectivity has been taken to mean different things in different circumstances.[14] Rather than attempt a catalogue of these interpretations, one may begin by noting that objectivity has almost always been understood in opposition to the term "subjective." Furthermore, one may take subjectivity to denote a dependence on the experiences of "subjects," or particular human observers; this provides a useful foil for comparing views of objectivity. That is, one may speak of objectivity as opposing subjectivity either in its connection with particular observers or in its dependence on observers at all. In other words, something may be independent of a particular subject by relying on more than one subject or on none whatsoever.

This gives rise to two ways in which scientific representation might be objective; these may be summarized on a simple grid so that subjectivity, S, may be contrasted with two forms of objectivity, Obj_1 and Obj_2 (see Figure 3.1).[15] The first of these, Obj_1, involves representational generalizations upon which more than one observer agrees; this might also be called objectivity as multi-subjective agreement.[16] The second, Obj_2, involves an ontological claim for the existence of a physical system in the world, W, represented by idealized model O and mathematical model M over and above the experiences of any number of human observers.

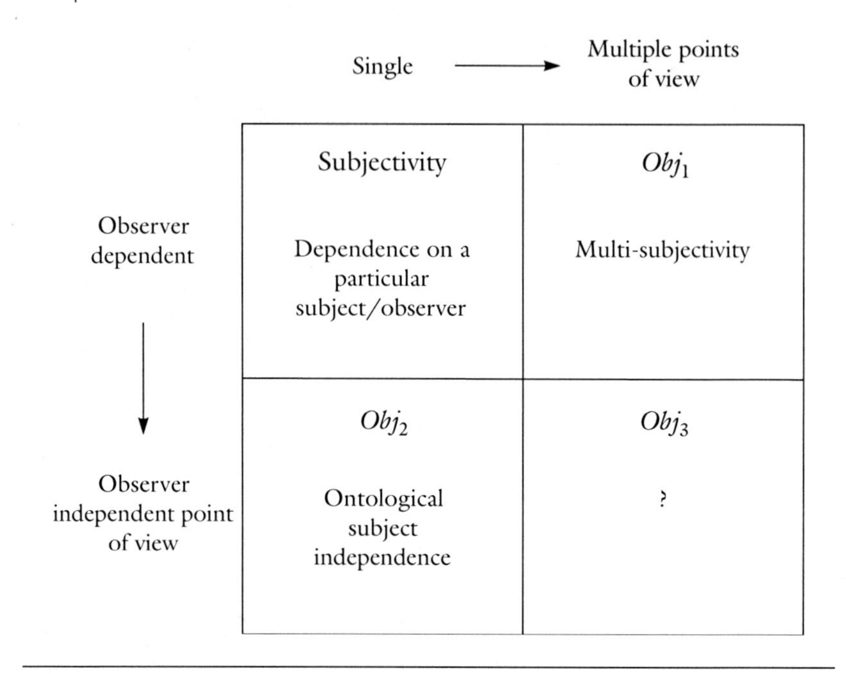

Figure 3.1. Opposites of subjectivity

If in using the slogan "objectivity means invariance" one intends objectivity in the sense of Obj_1, multi-subjective agreement, then one might be led to a view like that proposed by Bruno Latour. From his sociological perspective on science, he suggests that "Each scientific discipline . . . can be described by the choices it makes in what should be kept constant through what sort of transformations into different media."[17] Use of the term "transformation" here, as a shift between distinct physical media (measuring device to graph paper, for instance), is distinct from the formal sense of transformation as a map between elements of mathematical structures. However, the two meanings are parallel in the sense that they both display what has been called the duality between change and permanence. For Latour, that which survives a transformation he calls an "immutable mobile."[18] Speaking in more familiar terms, an immutable mobile is just a structure preserved through the various physical and virtual manipulations found in scientific practice. Crucially for Latour, this structure is preserved across transformations, which include the social practice of transferring information.

This view is undoubtedly accurate in the obvious sense that scientists do indeed transfer information about structures in such a way that this information is preserved along a social chain of media, as discussed in Chapter

1. Indeed, this is characteristic of the social dimension of representation. However, this sociological sense of invariance does not seem to give much substance to the slogan that objectivity means invariance, since information of all kinds—fact, fiction, or fantasy—can be faithfully preserved from one social agent to the next.[19] Thus, invariance in the social dimension of scientific representation is, on its own, too general to be a serious candidate for capturing the full significance of symmetry in modern physics. This is not likely to concern Latour very much, who has other philosophical considerations in mind. Nevertheless, this sociological notion of invariance captures an important, if general, aspect of scientific representation in practice.

Alternatively, if in using the slogan "objectivity means invariance" one intends objectivity in the sense of Obj_2, ontological subject independence, then one can still hope to capture this sociological multi-subjective dimension. Indeed, within the study of science, there is a strong tradition maintaining that subjects agree because there is a fact that exists outside of human experience. Thus, the multi-subjectivity of Obj_1 is precisely what leads one to hypothesize the ontological claim of Obj_2.

Following this tradition, it would appear difficult to maintain Obj_1 and Obj_2 as distinct positions. This difficulty may be addressed by noting that in Obj_1, multi-subjective agreement, by the very notion of agreement on something, one is tempted to think in terms of abstracting from particulars; i.e., something (in this case a model or other structure) that is distinct from the subjects who agree on it. One may conclude, as already suggested, that this abstract entity corresponds to a fact, but one may also take a nominalist approach by denying the existence of any abstract entity beyond the two subjects in the first place. In other words, one may be a working realist with respect to scientific models in much the same way as to mathematical entities, as previously discussed. One may thus maintain a distinction between two notions of objectivity, Obj_1 and Obj_2, corresponding to multi-subjectivity and ontological subject independence respectively.

The grid constructed in Figure 3.1 also has a conspicuous blank; one might wonder whether this could be filled by a third distinct notion of objectivity, Obj_3. If so, then it should combine the move from single to multiple points of view taken when opposing *S* with Obj_1 and the claim of ontological subject independence indicated when opposing *S* with Obj_2. Weyl's notion of objectivity as group-theoretical invariance is situated within this third category, Obj_3. Indeed, this is the sense in which proponents of invariantism generally use the term "objective" in reference to group-theoretical invariance. To see that this is the case, one needs to take a closer look at what is claimed by proponents of this approach to account for the philosophical significance of symmetry.

Invariantism

Pointing out the inadequacy of invariantism has not been systematically attempted in the literature. Indeed, the term itself is being proposed here. Before mounting a critique, however, one needs a more complete characterization of its central claims. At the outset it should be noted that this characterization represents a relatively broad philosophical approach. There are, nevertheless, some recognizable features that warrant coining the new term. First among these is a condition for objectivity based on meeting a criterion of invariance under certain symmetry transformations. Starting from Weyl's slogan that "objectivity means invariance with respect to the group of automorphisms,"[20] this might be taken to mean that: "Invariance with respect to the group of automorphisms is both necessary and sufficient for objectivity." Second, there is a recognition of the power of the heuristic method described above, and an attempt to establish a connection between the proposed condition for objectivity and this heuristic success. Third, there is an attempt to establish a connection between the condition for objectivity and universality. Such are the central features of an approach that many physicists and philosophers claim as their starting point.

Weyl's claim is the simplest statement of this approach. In what sense might he have intended that an invariant representational structure be considered objective?[21] Citing one of his own examples, the claim that there is a vertical direction, indicated by a vector, fails to be objective because of the fact that gravity defines different vertical directions at a different locations—for example at one of Earth's poles as compared to at its equator.[22] In other words, the concept of a vertical direction (as defined by gravitational fields) is not invariant under changes of location, or spatial translations, and thus according to Weyl fails to be objective. On this account, then, invariance is a necessary condition for objectivity. In addition, behind the slogan that "objectivity means invariance" is the intuitive observation that things which appear the same from different perspectives are objective. How this intuition makes invariance a sufficient condition for objectivity will be discussed in detail later. For now, it will suffice to point out that invariantism proposes that invariance under a specified group of automorphisms is both a necessary and sufficient condition for objectivity.

What also becomes evident is that for many invariantists like Weyl, the symmetries of nature are understood as the group of automorphisms invariance under which implies objectivity. Weyl is well aware that not all phenomena respect the "full symmetry of the universal laws of nature," which he takes as evidence that some physical phenomena are simply contingent facts about the world.[23] Even so, as this quotation reveals, Weyl shares the perspective expressed by many physicists that there exist a special set of symmetries of nature and that these are related to universality in scientific representation. Moreover, these symmetries of nature are effectively just those symmetries that are most relevant to the leapfrog heuristic described above. Thus Weyl's position exemplifies all three of the central features of invariantism suggested above.

Weyl's approach to symmetry may be understood in more general terms as a way of expressing the traditional intuition that theory should hold and experiments must be repeatable under all possible conditions. For the vertical vector, this means that it should point in the same direction no matter where it is situated in space; one might imagine a little experiment with a plumb line being repeated at different locations. This way of approaching objectivity seeks to move away from the particular (each little experiment) to generalizations of the experiences of individual observers, a move that follows the spirit of Obj_1, or objectivity as multi-subjectivity. In addition, for Weyl the symmetries of nature, in this case including spatial translations, exist as entirely independent from human subjects; this implies the ontological claims of Obj_2, objectivity as ontological subject independence. Thus Weyl's position seems to exemplify the third category, Obj_3, which combines the two notions Obj_1 and Obj_2.

More recently, a more thoroughgoing philosophical version of invariantism has been proposed by Robert Nozick. This version is ultimately intended to apply not only to scientific representation but much more broadly to any statement of fact. He claims, "The special necessary truth about objectiveness, the one that it exhibits in all possible worlds, is the *broad* notion of being *invariant under all admissible transformations.*"[24] This statement is intended to capture the invariantist intuition that objective facts are those which are invariant under some group of transformations; in the form of a condition for objectivity this indicates once again that invariance implies objectivity.

Nozick recognizes the need to specify which group. For this reason he adds, "The *deep* notion of objectiveness (for a specified world) is that of being *invariant under specified transformations* . . . [known only] . . . through the bootstrap process of scientific investigation."[25] Nozick's bootstrap process is very similar to the leapfrog approach described above. It differs in one crucial aspect. For Nozick, the question of which group of automorphisms to be specified is determined by a process akin to, and possibly ultimately identical with, biological natural selection.[26] Thus in this case, the heuristic application of symmetries to scientific models effectively determines the group of automorphisms under which invariance is ultimately required. In this sense the group of symmetries that constitute the condition for objectivity is determined by the heuristic success of those symmetries. Thus, objective status is not conferred on a model because it is invariant under a fixed and ideal set of symmetries of nature, as appears to be the case for invariantists like Weyl.

This suggests that Nozick's working concept of objectivity is slightly more subtle than Weyl's. Indeed, Nozick advocates what he suggests might be called "objectivity-at-a-level."[27] By this he intends for there to be different degrees of objectivity that vary roughly with the size of the group of symmetries under which invariance is required. This he intends to be a partial ordering of objectiveness, since the groups under consideration may often be disjoint. In this way, Nozick is not tied to a single group of symmetries of nature as is Weyl. However, Nozick's view is not without metaphysical implications. It is, after all, heuristic success in the bootstrap process of scientific investigation that leads one to consider the invariants of a certain set of symmetries as objective at some level. This success may indicate that there is something ontologically subject-independent about the significance of these symmetries. Moreover, he indicates that objectivity in the sense of independence from the knower should depend at least in part on invariance features.[28]

Thus, whereas Weyl depends on a single metaphysically significant group of symmetries to underpin objectivity, Nozick relies on numerous metaphysically significant groups of symmetries that help to confer lesser or greater degrees of objectivity. This subtlety notwithstanding, Nozick's proposal bears all the hallmarks of invariantism. First among these is a condition for objectivity based on symmetry or invariance, as expressed above. Second, he sees a connection between objectivity so defined and the heuristic success of symmetry in science; in fact, his conviction that physics has demonstrated this connection is part of his motivation for trying to apply it to other realms of discourse.[29] Finally, he sees a connection between objectivity and universality, this time in the sense of universality at a level depending on the symmetries under consideration; he thus suggests that it may turn out that there are no statements of the greatest level of objectivity, those "whose truth is invariant across all possible worlds."[30] This pessimistic attitude toward metaphysical necessity is reminiscent of van Fraassen's rejection of necessity in scientific laws, but it is significant that Nozick's notion of objectivity-at-a-level still leaves room for what might be called universality at a level.

A Sufficient Condition

The positions articulated above should provide a good sense of what is meant by the term invariantism. The popularity of views like these among philosophers and philosophically minded physicists has coincided with the tremendous practical heuristic effectiveness of invariance principles in modern physics. It will not have escaped notice that no argument has been provided to back up the claims made by Weyl and Nozick. Unfortunately, the connection between invariance and objectivity is all too often simply assumed to apply more or less directly, and then used by physicists and philosophers to reach even more general conclusions. Surprisingly, an explicit argument is not normally given to back up the central claim that objectivity means invariance, or more to the point that invariance is both a necessary and sufficient condition for objectivity.

An important starting point in addressing the supposed philosophical significance of invariantism is to provide such an argument. In this section a sufficient condition for objectivity in scientific representation will be argued for. Filling out the invariantist position in this way will provide a basis for criticizing it. However, the task here is not simply to provide a straw man. In fact, the sufficient condition proposed will also provide a positive aspect of the new appraisal of symmetry in physics presented here.

In order to accomplish all of this, one may begin with the intuitive notion that an objective fact is the same when viewed from any possible perspective. This intuition is common to the different senses of objectivity previously discussed (and summarized in Figure 3.1). By "perspective" one might mean a literal vantage point or simply the way a given observer is situated in some figurative sense. Thus, the number of Jupiter's moons will not change depending on whether it is observed from Los Alamos, New Mexico or Mauna Kea, Hawaii. Similarly, the acceptance of the objective fact that George Bush is a Republican will not vary depending on whether one has a liberal or conservative perspective on American politics. These basic examples illustrate the intuitive appeal of the notion that:

(A) Objective facts are those that are the same from any perspective.

Another way of saying the same thing is that objective facts are *invariant* under changes of perspective. At this level, there appears to be a strong affinity between the notions of objectivity and invariance.

Starting from the intuition (A) above, one may add that physicists typically view phenomena by representing them. Thus, one may propose a sufficient condition for objectivity:

(B) If a fact appears the same in any representation, then it is objective.

Here, one compares representations of the fact in question instead of perspectives on it. The motivation behind this sufficient condition for objectivity is still compelling. Intuitively, representations and perspectives relate to one another in a very natural way. If a perspective really is a vantage point on a phenomenon, then one may always choose to represent the phenomenon from that perspective. In reality, as previously suggested, the notion of a perspective in the strict sense of a spatio-temporal vantage point applies in only a limited number of cases within modern physical science. By shifting to the term "representation" here, one may actually consider a more general class of avenues of access to a given phenomenon.[31]

In order to make explicit the notion of invariance implied in (B), one must note a basic point that an entity is never simply invariant, but is instead invariant with respect to something which changes. Indeed, this exemplifies a general philosophical principle—the duality between change and permanence. With this in mind, sufficient condition (B) may be revised to include explicit reference to invariance:

(C) If a fact is invariant under all changes of representation, then it is objective.

Speaking in terms of a change of representation makes explicit the shifting among different perspectives implicit at the start in (A). Intuitively, objective facts should not change depending on the way they are represented. Now, one may claim that objectivity depends on *invariance* under certain specified changes, namely those between possible representations of O.

How might this characterization of objectivity be applicable to scientific representation? In the ideal case, as previously described, mathematical model M is isomorphic to O, an idealized conceptual model. Now, applying the proposed sufficient condition above, one may look for objective facts about *O* as represented by *M*. According to (C) above, these facts are objective if they are invariant under changes of representation; in this case, changes in model *M*.

But what changes in model *M* are possible here? There are in principle any number of mappings that specify changes in *M*. Among these, the automorphisms of *M* are transformations that map *M* back into itself. Since, in this framework, *M* is isomorphic to *O*, the only possible changes in *M* that also constitute changes in the way it represents *O* are those transformations which preserve its structure, its automorphisms.

The automorphisms of any structure form a set with special properties called a group. This group of symmetry transformations may be said to contain all of the possible changes that *M* may undergo in representing *O*. Furthermore, as *O* and *M* are being considered as isomorphic, the automorphism groups of both structures are abstractly the same group. One may thus rewrite the sufficient condition for objectivity in the case that *M* represents *O*:

> (D) If a fact about *O* is invariantly represented under the automorphism group of *M*, then it is objective.

Since *M* is taken to be isomorphic to *O*, then every feature of *M* represents a feature of *O*, and their respective symmetry groups, as stressed above, will also be isomorphic (as representations of the same abstract group). Thus, one may conclude that:

> (E) If a feature of *O* is invariant under its automorphism group, then it is objective.

This would appear to provide a sufficient condition for identifying objective facts—relational facts in a structuralist sense, not objective objects in the world. After all, in the framework envisioned here, the physical system, *W*, is only imperfectly represented by the idealized model *O*. This sufficient condition for objectivity can be seen as a formalization of the initial intuition about objectivity and perspectives (A) applied to the case of representing physical systems. Although similar claims are often made, arguments of this sort are not usually made explicit. It is not being suggested here that (E) represents exactly the same sufficient condition that proponents of invariantism like Weyl or Nozick propose, although it may be that (E) brings one closer to these than any argument they have provided. Since invariantism is based on some condition for objectivity based on group theoretical invariance, (E) is the most plausible candidate for which an argument is available.

Considering (E) in more detail, recall that in practice one may either start with the invariants of a structure and find its symmetries or the reverse; this follows from the duality between the notions of symmetry and invariance previously discussed. In the case that the physicist begins with some objective features of *O* and finds its symmetries, i.e., transformations under which these objective features are invariant, then (E) is merely tautologous. Knowing the truth of the antecedent of this conditional depends on knowing what the symmetries are. However, one might go ahead and determine the symmetries of *O* and then apply (E), after the fact, to identify the objective features of *O*. Then this would be a trivial application of the sufficient condition for objectivity, since the objective features so identified will be just those objective features with which the physicist began!

In the alternative case, however, one begins with just symmetries. In this situation, (E) functions straightforwardly as a sufficient condition for identifying the objective features of *O*, i.e., the invariants. Thus, as one moves from these symmetries to discovering the invariant features of *O*, one is also able to identify them as objective.

One may note that in this case there may be objective features of *O* that are not shown to be so by application of condition (E). These could be the identity of the very elements of structure *O* between which relations are specified. Obviously, the identity of these elements of *O* can't be invariant, since automorphisms of *O* carry one element into another. If one understands these elements as objects with objective identities, they cannot be identified by (E). Thus one sees that invariance is a sufficient condition but may not be a necessary condition for objectivity, where objectivity is understood as independence from perspective, as in (A) above. The fact that invariance is not seen here as a necessary condition for objectivity should be a vital clue that invariantism, which looks for just this, is not a supportable view.

Against Invariantism

It has been suggested above that invariantism has three inter-related central claims: that invariance is necessary and sufficient for objectivity, and that this objectivity may be connected with the heuristic power of symmetry on the one hand and representational universality on the other. It is not difficult to see suggestive and tantalizing connections between the concepts of objectivity and the discovery of new and more universal models. What was missing was an argument in support of the first definitive feature of invariantism, a condition for objectivity based on invariance. Having provided an argument in support of sufficient condition (E), one can see that without some additional assumptions, it is not possible to make invariance a necessary condition for objectivity. Thus, one can begin to see the weakness of the invariantist position.

Objectivity and Heuristics

It must be remembered that the heuristic leapfrog method is decidedly fallible, the most famous example of this being the violation of symmetry under spatial inversion (changes in parity) in the physics of the weak interaction. Regardless of the heuristic success of the leapfrog method, it is clear that not all symmetries are even intended to play the same heuristic role. What distinguishes certain symmetries from others is precisely their heuristic applicability, the extent to which they may be used to aid the process of discovering new and better scientific models. In fact, many symmetries that appear in the construction of scientific representations are not at all heuristically fertile.

Following a scheme first proposed by one of the authors, the symmetries associated with a scientific representation may be broken down into useful categories.[32] The largest category includes all of the automorphisms of the larger mathematical structure M', model M plus its surplus structure. These may be termed "mathematical symmetries."

The second category, a subset of the first, may be termed "physical symmetries" and includes the automorphisms of idealized model O. In the ideal case considered above, in which M is isomorphic to O, the automorphism group of M also represents the "physical symmetries" of O. The relative complement of these physical symmetries with respect to the mathematical symmetries is the set of *purely* mathematical symmetries.

The heuristic leapfrog begins with the notion that a certain set of symmetries are relevant to a particular physical system; from this one may then seek to generalize the applicability of these symmetries to one or more additional physical systems. Physical symmetries, however, do not all have heuristic potential. In fact, a number of these symmetries are entirely accidental. As one of the authors has pointed out, "the distinction between heuristic and accidental symmetries is not a categorical one," in that symmetries which appear accidental may later turn out to have heuristic potential; alternatively, symmetries which appear to have heuristic potential may later "be downgraded to accidental status."[33]

There are several distinct classes of heuristic physical symmetry as they have appeared in the history of the physical sciences.[34] The first type of symmetry is that which relates to descriptions of the same event at different spatio-temporal locations—these include especially the spatiotemporal symmetries. The second type relates to descriptions of different events at the same location—these include charge conjugation, permutation symmetries, and SU(3) symmetry of particle physics. The third type of heuristic physical symmetry relates to descriptions of different events at a different location. The fourth type of heuristic symmetry relates to the redescription of the same event at the same location—this includes, for example, gauge transformations. In this case these are only trivially physical symmetries, but perhaps surprisingly they do turn out to have heuristic potential.[35]

Finally, one may make one more distinction between universal and dynamical symmetries. It is claimed that some heuristic physical symmetries are universally applicable as opposed to being simply very widely applicable. It may be questioned whether one can even know if symmetries of this type actually exist—especially since the so-called "symmetries of nature" are always, in principle, empirically revisable. Those physical symmetries that are not universal are often called dynamical, since they relate to special sorts of interaction, and may include heuristic symmetries and accidental symmetries.

Although this classification of types of symmetry is undoubtedly helpful, one can note that there is no account given for why certain symmetries turn out to be universally applicable, others dynamical with wide heuristic application, others dynamical but purely accidental, and still others purely mathematical and associated with surplus structure. In fact, history seems to suggest that no general account will emerge. It is important to note that the significance of certain symmetries and the insignificance of others seem to be a brute fact about the way physical science represents the world.

The Definitive Group

The second criticism of invariantism is over this question of which group of automorphisms to take as definitive. For Weyl, determining which symmetries count as the symmetries of nature is a major difficulty in making sense of his claim. In reality, the symmetries of nature do not represent a single, even an ideal, group of symmetry transformations. This is due to the fact that different aspects of the physical world have different symmetries. Take for example two physical systems, the hydrogen atom and special relativistic spacetime, represented by different idealized models, O_1 and O_2 respectively. These two structures have different symmetry groups, the $O(4)$ dynamic symmetries of the hydrogen atom and the spacetime symmetries of the Poincaré group, respectively. Crucially, in the latter case the symmetries of O_2 result in a heuristic leapfrog from one domain of inquiry to another.

But in the former case the symmetry does not generalize; it does nevertheless serve a very important function in identifying the degeneracies in the hydrogen spectrum. Ultimately, the test is case by case and empirical. Thus, physicists have discovered that the spacetime symmetries of special relativity theory may be applied to representations of many different physical systems, but that the dynamic symmetries of the hydrogen atom may not. Since different physical systems have different symmetries, Weyl's attempt to ground objectivity in the symmetries of nature must reduce to a claim about what have been called the universal physical symmetries, above. One problem for Weyl is how to determine which symmetries constitute this very special set, apart from their heuristic fertility. If these symmetries are so selected simply due to their heuristic effectiveness, then why add to this the notion that they are associated with objectivity?

Nozick's invariantism has similar difficulties. Nozick contends that the choice of which group is relevant comes down to those which have proven themselves most effective in the bootstrap process of scientific investigation. This response amounts to relying on the heuristic application of symmetry methods as the basis for defining objectivity. This does not provide an account of objectivity which is based independently on the intuition that objective facts are those which are the same in any representation.

But this is only part of the story as, on Weyl's account, it is invariance under a single specific group of automorphisms that is the crucial determinant of objectivity, spatial translations being only a few of its members. Indeed, he warns, "Reality may not always give a clear answer to the question of what the actual group of automorphisms is."[36] However, even if one insists that this problem can be surmounted, that a definitive group of symmetries of nature can be found, then another difficulty arises. Returning to Weyl's example of a vector that defines the vertical direction, if this vector must be parallel to the local gravitational field, then no single direction seems to be invariant under spatial translations.[37] Following his logic, since spatial translations are members of the group of automorphisms of nature, then there is no objective vertical direction.

However, one might save the objectivity of the vertical by adopting a simple relational strategy. This may be done by re-defining vertical as vertical in relation to a given position. Thus a vector, V, which is vertical at position (x, y, z) may be moved to a new position at which it fails to be vertical (i.e., parallel to the local gravitational field), but it is still the case that V is vertical when at position (x, y, z). Thus the new definition of vertical is invariant under spatial translations and qualifies as objective in Weyl's sense. This example suggests that theoretical terms, like vertical, may be made objective (with respect to a definitive group of transformations) by adopting a relational strategy that makes the modified term invariant.

This sort of relational objectivity, its potentially distasteful *ad hoc* nature notwithstanding, satisfies Weyl's definition of objectivity perfectly well. However, a problem arises for those committed, as he is, to a definitive group of symmetries because the strategy of redefining the term "vertical" in relation to spatial positions is equivalent to removing spatial translations from this group. It is simply not possible to move a vector V at (x, y, z) by a spatial translation with respect to the coordinates (x, y, z); it has been pinned down to this location by our relational strategy. This strategy has in effect brought about a shift to a different group of transformations by excluding spatial translations. This in effect violates the assumption that the original group was *the definitive* such group of automorphisms. Thus, even with a determination to reserve one group as the actual or potential symmetries of nature, one may effectively use different groups as a basis for objectivity.

Objectivity and Universality

The assumption that there is a definitive group of symmetries is also made in the attempt to use symmetry considerations to arrive at a universal theory of physical phenomena: only the symmetries associated with such a final *universal* theory would also be the correct symmetries of nature. This intuitive assumption underlies much talk of discovering the symmetries of nature; i.e., why discover them unless they are out there to be discovered? If symmetry has been thought fundamental to physical science as a means to universality and as a means to objectivity, both seem to depend on the assumption that there exists a set consisting of the symmetries of nature and to push physicists in the direction of finding them. Thus the search for objectivity, in Weyl's terminology, may be linked closely with the search for increasingly unified theories. This link only exists with reference to certain symmetries.

This link, and in particular the assumption that underlies it, presents a problem for using symmetry as a means to objectivity. The primary difficulty for objectivity defined, following Weyl, as invariance with respect to the symmetries of nature, is the implication that the determinations of physical theory will never be fully objective until the true group of automorphisms is known. For example, the use of classical physics to represent space and time must fail in its *objectivity* because its models are not invariant under the spatio-temporal symmetries of relativity theory. It is one thing for classical physics to fail to be accurate in certain regimes, perhaps those in which very large velocities are involved, but it seems overly restrictive to conclude that its representations fail in terms of their objectivity. In other words, one would like to retain the capacity to make objective claims that are nonetheless provisional. Furthermore, unification is a project with a different end; one may expect not to reach final unification until one gets there. Objectivity, however, is something one would like to have regardless of the success or failure of the effort to find a unified final theory.

Perhaps in defense of Weyl, one might object that objectivity should indeed be related to a unified final theory. If objectivity intuitively involves a sense of subject-independence, (i.e., the opposite of subjectivity), then what could be more so than a final, correct theory of all phenomena? The proponent of this view would, in other words, be willing to forego objectivity as a possibility in the present and see it instead as an ideal to be attained alongside the discovery of a final theory. This may not be surprising since objectivity as universalized subject-independence always ends up functioning as an ideal within the context of particular human subjects and their practices. However, if a definition of objectivity in terms of symmetry is to be of value, quite apart from the search for a unified theory of everything, then it should offer more than this.

A Tantalizing Illusion

On closer inspection, invariantism, which has captured the imagination of so many, may not be a viable approach. The arguments provided above were intended to provide a reasonable account of this philosophical approach to the physical sciences. Although not often argued for in all its detail, this approach appears frequently in the philosophical literature and in the statements of working physicists. Invariantism captures an approach to understanding invariance in the sciences that sees a link between objectivity and invariance and finds support for this view in the remarkable heuristic success of symmetry methods in physics. Much confusion over the evaluation of the claims of invariantism has been caused by the fact that clear arguments are not usually given for the connection between objectivity and invariance, and the claimed role of such objectivity in accounting for the success of symmetry methods in physics.

The significance of this appraisal of invariance in physics is brought into relief by understanding the attraction of invariantism in the first place. If invariance under a group of automorphisms were both necessary and sufficient for the objectivity of model *O*, then one might be able to test various models for their objectivity in a more or less algorithmic fashion. One wants objective models largely because objectivity is often thought to imply ontological independence. If this were the case, the physicist in search of mind-independent truth would have in invariantism a method with a very significant payoff.

This is made all the more alluring since physicists often seem to use invariance principles in just such a manner, through the leapfrog heuristic, which in turn seems to provide further support for the philosophical significance of symmetry. Success of invariance principles as a heuristic tool in physics is indisputable. That this success can in general be explained by an invariantist link to objectivity, sadly, is not.

Perspectival Invariantism

As demonstrated above, invariantism in general is a failed philosophical approach to understanding the fundamental significance of symmetry in physics. However, a limited form of invariantism can be introduced when considering certain specified symmetries, especially those of space and time. This depends initially on the fact that a connection between symmetry transformations and changes of perspective is plausible under these limited conditions. These cases are important on the one hand as a way of accounting for part of the intuitive appeal of the general invariantist view and on the other as a means of illustrating the fact that even where invariance might imply objectivity, it also implicates convention.

A Necessary Condition

As previously discussed in detail, invariantism has three inter-related central claims: that invariance is necessary and sufficient for objectivity, and that this objectivity may be connected with the heuristic power of symmetry on the one hand and representational universality on the other. The perspectival invariantist claim is that there are important cases for which these claims hold, when the symmetries involved are appropriately specified. In order for this to be so, one must overcome a limitation of the argument provided above in support of invariantism. Namely, that invariance was shown to be merely a sufficient condition for objectivity.

This is due to the simple fact that, as in this example, there may be objective features of some idealized model *O* which are not shown to be so by virtue of invariance. A more basic example might be the identity of the very elements of structure *O*. As previously explained, the identity of these elements of *O* can't be invariant if automorphisms may be said to carry one element into another. Therefore, if one understands these elements as objects with objective identities, they cannot be identified by virtue of their invariance and one must conclude that invariance is a sufficient but not necessary condition for objectivity. Overcoming this represents the final obstacle to constructing a workable perspectival invariantism.

A pragmatic response is to find a way to define objectivity so that invariance can be seen as necessary as well as sufficient. In order to do so, one must be willing to accept that the identities of elements of an idealized model *O* will fail to be objective in this new sense of the term. In other words, the kinds of features of *O* that are substantially objective are relational features of model *O*. As such, these features characterize an abstract structure in the sense discussed previously. Taking this step to restrict objectivity to the relational features of *O* is within the spirit of structural realism, which sees abstract structure as especially significant in the sciences. Structural realists may see this significance as epistemological or as extending to include ontology as well.[38] Thus the structural realist might agree with James Ladyman in the claim that "theories tell us not about the *objects* and *properties* of which the world is made, but directly about *structure* and *relations*."[39] The structure about which one might be a realist here is necessarily abstract. One of the authors has claimed, "The aspect of reality that is amenable to scientific investigation is the abstract structure as represented for us by the mathematical structures of theoretical physics."[40] This is very close to the spirit of the approach advocated thus far, according to which the question of the world out there has been bracketed, and a formal scheme comprised of an object in the world, *W*, an idealized model, *O*, and a mathematical model, *M*, has been adopted. This is of course no accident since the dual account of scientific representation advocated in Chapter 1 explicitly sought to conceptualize the formal chain of media (*M* and *O* taken together) in terms of structures.

Thus, it will be consistent to take the step of limiting the objective features of idealized model *O* to those that are relational features of *O*. Incidentally, the argument provided here is not unlike that provided by John Earman in reference to the relationship between dynamical and spacetime symmetries. In particular, he rules out ascribing rigid designators to regions or elements of spacetime in defense of the proposal that "Any spacetime symmetry is also a dynamical symmetry."[41] In any case, designating objective features in this way will allow for making invariance both necessary and sufficient for objectivity, and thus open the door to consider cases which might legitimately exemplify perspectival invariantism.

In short, on this view the objectivity of a feature of *O* is equivalent to its invariance under the automorphisms of *O*, when these automorphisms may be interpreted perspectivally and are generalizable (and therefore heuristically fruitful). Moreover, this also implies an understanding of representation that restricts discussion to relational features of *O*. Here there are two underlying intuitions about objectivity. The first is the notion that objective facts are the same from any perspective. The second is the notion that objective facts are restricted to the relational features of scientific models which survive transformation from one model to another. The type of objectivity implied by a perspectival invariantist approach might be called perspectival objectivity (Obj_P).

As with invariantism in general, to use perspectival invariantism to identify objective features of some idealized model *O*, one must begin by specifying a group of automorphisms. Since in this case it is claimed that invariance may be considered both necessary and sufficient for objectivity (Obj_P), one may conclude that:

> (F) If a feature of *O* is not invariant under the specified automorphism group, then it is not objective (Obj_P).

Although not generally applicable, as they appear to have intended, this may be the closest supportable rendering of the Weyl-Nozick invariance condition. Furthermore, given a specific relational feature of *O*, one might refer to the group here specified as an invariance criterion. In practice, one will actually consider the invariance of the corresponding relational feature of mathematical model *M* under a group of automorphisms that is a representation of the same abstract group as the automorphisms of idealized model *O*.[42]

This chapter has presented two readings of the view here termed invariantism. The first argues against invariantism as a general account of the philosophical significance of symmetry in physical science. The second reading seeks to suggest an adequate mitigating view that allows for a limited form of perspectival invariance. These readings parallel two sorts of philosophical approach relevant to this question. First of all, one may adopt a realist construal which maintains that symmetry arguments are external to the dynamics. There are facts one can never know, but one can apply the Rule of the Excluded Middle to argue that they must have a truth value that has nothing to do with one's ability to know what such value may be. Alternatively, one may adopt a positivist approach, according to which one can only know what can be inferred from empirical fact; this leads directly to the Weyl-Nozick invariance requirement and to Earman's view that symmetry is dictated by the internal dynamics of a theory. One might argue from realist intentions for the first reading of invariance, but the legitimacy of the second must also be recognized in the notion of perspectival invariance. Indeed, the following case studies are pursued from this latter point of view.

4

Simultaneity and Convention

The first of three case studies to be considered also evokes one of the best-known philosophical controversies concerning conventionalism in modern physical science. The determination of synchrony, or simultaneity, between distant clocks within Einstein's special relativity has long been cited as an example of irreducible conventionality in physics. Those who have argued for the conventionality of simultaneity have been strenuously resisted by proponents of the relativity of simultaneity as a basic, if counterintuitive, consequence of special relativity theory. Although many have assumed this question settled in favor of the relativity of simultaneity, due to a proof given by David Malament, this chapter will illustrate that this is not at all the case. In reality, the formalism of special relativity seems to allow for both accounts, and this formal equivalence will be made explicit. There is of course still a difference between the views, but it will be shown that from a perspectival invariantist perspective this difference may be best understood in terms of what structures are held to be objective *(Obj_P)* in the sense of being invariant under a group of automorphisms, an invariance criterion. In these terms, the question of objective simultaneity is coextensive with the sense in which it is a conventional relation.

Simultaneity

The significance of simultaneity, the notion that two distant events happen at the same time, finds its early modern roots in Newtonian physical theory. The view that emerged from Newton's mechanics, as commonly understood, treats space and time as an absolute and unchanging background for the interaction of objects and forces. One feature of this paradigm was the proposition that time proceeded uniformly across all of space; or equivalently that a given moment on Earth could be unambiguously associated with the time reading on a hypothetical clock anywhere else in the physical universe. Newton's dynamics proposed that forces could act instantaneously across vast astronomical distances; this famously left the theory open to philosophical criticism of the apparent "action-at-a-distance."

Einstein's special theory of relativity, famously introduced in 1905, was in many ways strikingly different from the Newtonian conception that had dominated physical theory for roughly two hundred years. Central to special relativity was a novel concept of distant simultaneity. In fact, Einstein began by visualizing three-dimensional space with clocks positioned periodically along this lattice of coordinates.[1] As the new theory was elaborated by H. Minkowski and others, one could still embed simultaneity assumptions in the way one chose to coordinatize the newly interdependent spacetime manifold. On this so-called Minkowski spacetime, synchrony between events can be indicated by hyperplanes of simultaneity.[2] The hyperplane, or simply plane, of simultaneity for a resting inertial frame bears a strong resemblance to the sort of relations one could construct using Newtonian assumptions about space and time. The difference is most clearly apparent when one considers the fact that under the new theory each moving frame of reference has its characteristic plane of simultaneity, different from that of the resting frame. This feature of special relativity has typically been called the relativity of simultaneity. Philosophically this state of affairs undermines theories of time that view the present as a single, simultaneous boundary between past and future, a universal horizon of temporal becoming.

Finally, there is at least one meaningful general sense in which simultaneity, as it is used outside of the physical sciences, is significant to understanding scientific representation. This is little more than the notion that certain entities, structures, or events exist or take place all at once. From the perspective of the more precise formulations of modern physical theory, this may or may not imply nonlocality, or the existence of infinitely fast causal influences (which will be considered in some detail in Chapter 6). Typically, however, one would like to distinguish between actions or events that take place diachronically, forming a history or narrative, and those that take place synchronically. A diachronic sequence is usually contrasted to an entity that is supposed to exist outside of time or to transcend temporal particularity, a snapshot of reality. To the extent that all scientific representation is situated in space and time, one's general intuitions about simultaneity and the distinction between synchronous and diachronic will affect the way in which scientific models are interpreted. This is yet another reason it is especially important to address the topic of simultaneity as it appears in modern physics.

The Conventionality of Simultaneity

Hans Reichenbach introduced the problem of simultaneity as that of the "third coordinative definition of time," the first two being the definition of a unit of time and the assertion that successive time intervals using this unit are of the same magnitude.[3] Together, these give one the ability to measure and compare time intervals at a given position in space. Simultaneity is accordingly the convention whereby one may compare such time intervals measured at different spatial positions. More often, simultaneity is framed with respect to the problem of synchronizing two distant clocks.

This is of course precisely the situation that Einstein considers when laying out the principles of his special theory of relativity. Einstein suggests a particular operation that can be carried out to synchronize any two clocks at a known spatial distance from one another through the use of light signals. For two clocks at positions A and B, separated by distance d, this might consist of measuring the time interval between sending a signal from A to B and receiving the signal back again at A. Making the assumption, as Einstein does, that the speed of light is a constant, c, then the instant the signal is reflected from B must be simultaneous to the instant on the clock at A exactly halfway between the time of sending and receiving the signal (see Figure 4.1). That this is the case should be clear from the assumption that the light signal travels at a fixed speed in each direction, so that the time it takes to get from A to B should be equal to the time it takes to make the return journey. The operational character of this definition of simultaneity is thus apparent from the beginning.[4]

In considering Einstein's synchrony operation, Reichenbach made a few significant observations.[5] First of all, he noted that any signal, not just a light signal, may be used, as long as one knows that it is a "causal chain" of events; this philosophical emphasis on causal influences is worth noting in light of subsequent debate. Second, he pointed out that to speak in terms of the velocity of *anything* is to presuppose that synchrony has already been established between the initial and final location over which that velocity is measured. This observation is self-evident when one recalls the special relativistic claim that moving clocks do not retain absolute syn-

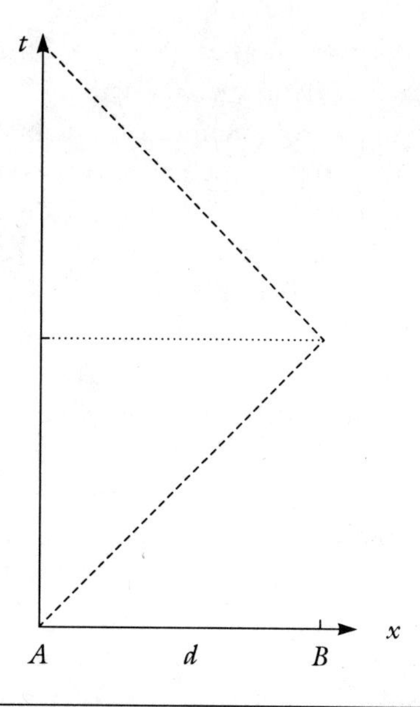

Figure 4.1. The synchronization procedure between two distant clocks at positions A and B. The dashed lines depict radio or light signals. The dotted line depicts Einstein synchrony.

chrony (see Chapter 5), so that one may only use stationary clocks to measure velocity in a given frame of reference. Taking one's commute to the office as an example, in order to calculate the average velocity for that trip (distance traveled per unit of time; miles/hour perhaps) one needs to measure a time of departure from home and a time of arrival at work, each on a clock held stationary at that location—no wristwatches allowed. For these two clock readings to be compared, in order to measure the elapsed time of the commute, they must have been previously synchronized. If therefore speaking of velocity implies prior synchronization of clocks, then Einstein's central special relativistic assumption of a fixed *velocity* for light would seem to do the same. But this is problematic in light of the fact that the basis for Einstein's synchronization procedure assumes a constant velocity c; thus his attempt to define synchrony in terms of light signals of fixed velocity seems to assume synchrony at the same time. In other words Einstein's definition of simultaneity is, in some sense, circular. On this basis, Reichenbach concluded that Einstein's proposed operation merely set the boundaries for an infinite number of possible definitions of synchrony.

This assertion in one form or another has formed the basis of the conventionality thesis on simultaneity in special relativity.

Returning to the example of commuting to work, if instead one measures the time it takes to go to work and return home immediately, then the moment of departure and return can be measured on the same clock at home. From this measurement, one can calculate the average velocity over the round-trip journey (twice the distance). This avoids the need to synchronize the clock at home with the clock at work and thus removes any prior assumption of synchrony, which leads to circularity when applied to Einstein's synchronization procedure. Furthermore, this measurement would provide an idea of what the commute time might be under the assumption that, for instance, the average velocity for the round trip was the same as that for the one-way trip. This is exactly analogous to the discussion of synchrony provided by proponents of the conventionality of simultaneity.

One may address this issue more formally in terms of some simple notation introduced by Reichenbach. This notation has become standard within the subsequent debate over the conventionality of simultaneity. Given clocks at positions *A* and *B* as before, one can measure the time of sending a signal, t_1, and receiving the reflected signal, t_3, on a clock at *A*. The interval between these two measurements provides a set of possible time values on the clock at *A*, which may be taken to be simultaneous to the value on the clock at *B* at the moment the signal is reflected back to *A*. If one calls this possible simultaneous value at *A*, t_2, then it may be expressed as follows:

$$t_2 = t_1 + \varepsilon(t_3 - t_1) \quad (4.1)$$

The epsilon, ε, term in this expression is a parameter that ranges from 0 to 1. This restriction has the essential function of preserving the relations among t_1, t_2, and t_3 such that t_2, the prospective simultaneous point, is between t_1 and t_3, the interval between sending and receiving the signal (see Figure 4.2). Any choice one makes for the value of the parameter ε produces a different value for t_2 within this interval. Each value of ε then establishes a distinct standard of simultaneity. It is the conventionalist thesis that there is no fact-of-the-matter at stake in asking which of these standards is the correct one.

If, for instance, one takes the value of $\varepsilon = 1/2$, then the expression above (4.1) becomes equivalent to Einstein's simultaneity criterion, constrained by the assumption that the signal from *A* to *B* must last as long as that

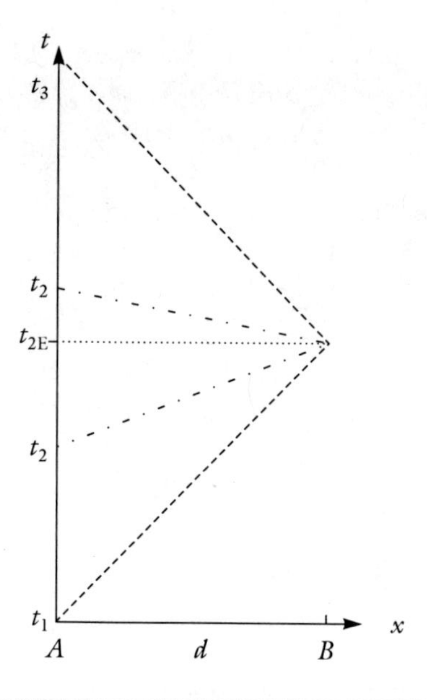

Figure 4.2. Conventional ambiguity in the synchronization procedure between distant clocks, at positions *A* and *B*. The dashed lines depict radio or light signals. The dotted line depicts Einstein synchrony. The mixed dashed lines depict possible nonstandard simultaneity conventions.

from *B* back to *A*. This is equivalent to, and indeed follows from, the assumption that the speed of light, *c*, is constant on outgoing and returning legs of the signal's round-trip journey. Conversely, when ε takes a value other than ½, this would seem to imply that the light signal takes a longer or shorter time to go from *A* to *B* than to return. Since the distance between the two is fixed, this implies that the speed of light is different on outbound and return legs.

This last implication, that the one-way velocity of light is not equal to the constant value, *c*, may seem distressing from a special relativistic perspective, the constancy of the speed of light being one of the two fundamental premises of the theory. The conventionality thesis does not, however, suggest a violation of relativity theory here, if one accepts the argument that one-way velocities, in order to be defined, require spatially separated synchronized clocks. Since this is actually one of the premises leading up to the conventionality thesis, it is perhaps surprising that its critics have tended to forget this and focus specifically on the conventionalist claim regarding the one-way speed of light in particular. As Wesley Salmon emphasizes, "*it is the very concept of one-way velocity whose status as conventional or non-conventional is at issue,* not just the one-way speed of light."[6]

According to the conventionalist, only the velocity of a signal over a round-trip path through spacetime avoids circularity, as the relevant time interval is measured on the same clock. In this way one can still assert that over the complete distance, $2d$, from A to B and back again, the light signal moves with a constant velocity. According to the conventionalist argument this is the real content of Einstein's claim that the speed of light, c, is a constant. This does not imply, as some might be tempted to think, that the speed of light depends on the direction of its path. The round-trip speed of light is still held to be constant regardless of the orientation of its path. That this must be the case should be clear from the fact that only total distance traveled and total elapsed time on a single clock are used to calculate c; the path *per se* does not enter the calculation.

The reader might object at this point that one-way velocity assumptions have been surreptitiously introduced into Figure 4.2 by depicting the velocity of a light signal as a straight line with a constant slope (using the standard $c = 1$ unit convention). After all this depicts a path through spacetime with a well-defined one-way velocity. This observation actually serves to further illustrate the point that having such a well-defined one-way velocity implies a prior definition of simultaneity. Before a worldline, a smooth curve in spacetime, can be depicted, one must have an inertial frame of reference. In this case, it is represented in a flat spacetime with one dimension each for time and space, known as $1 + 1$ dimensional Minkowski spacetime. Any set of Minkowski spacetime coordinates simply must provide a temporal and spatial axis, and these axes define lines of constant time and position respectively. Therefore, in order to depict the signaling scenario in Minkowski spacetime, a convention has already been chosen. This then becomes the basis for depicting the progress of a signal through spacetime in Figure 4.2. One should expect circularity in this case, since one is in effect using coordinates, the spatial and temporal axes of Figure 4.2, to depict a process by which those very coordinates, in this case lines of constant time, are supposedly being constructed.

Accepting the conventionality thesis as expressed in (4.1), one still gets the isotropy of the (round-trip) propagation of light, and all the major features of the special relativistic picture. This has been studied in detail by several authors, most famously by John Winnie, who derived the equivalent of the Lorentz transformations for unspecified values of ε.[7] He shows that a version of the Lorentz transformations may be used without choosing a single value for ε. However, he also points out that two of the best-known relativistic effects, the Lorentz length contraction and time dilation, are dependent on choice of simultaneity criterion. These kinematic effects will be addressed in more detail in the following chapter.

For Reichenbach the restriction on values of ε to the interval $[0,1]$ is significant because of its implications for what he proposes as the correct way to think about simultaneity, a notion he calls the "topological definition of simultaneity."[8] He proposes this definition as a new way of thinking about the relationship of simultaneity between points on a Minkowski spacetime manifold. Instead of adopting one of the infinite number of simultaneity conventions, each determined by a distinct value of ε in (4.1), one should in essence adopt them all. Since there is no fact-of-the-matter in choosing between these conventions, one may assert that all points t_2 in the interval $[t_1, t_3]$ measured at position A are simultaneous to the moment of signal reflection measured on the clock at B.

Reichenbach arrives at this solution by noting that the conventionality thesis (4.1) implies that there is no fixed time-ordering between t_2 and the reflection time measured at B. He asserts additionally that fixing time-ordering of events is an essential philosophical feature of causal chains of events and thereby of the notion of simultaneity. He expresses this in the claim that *"any two events which are indeterminate as to their time order may be called simultaneous,"* or alternatively that, "*Simultaneity* [between events] *means the exclusion of* [their] *causal connection.*"[9]

This is within the long-standing tradition in the philosophy of time to view the set of current simultaneous events as a moving present that divides time into past and future. Hermann Weyl, for example, takes precisely this approach to the light-cone structure in special relativity. He suggests that the light-cone, defined as the propagation of light in all directions from a given point on a worldline (or path through spacetime), *O*, divides spacetime into an "active future" and a "passive past."[10] Topological simultaneity does just this, precisely because its boundaries are set by just such a light-cone structure. Thus one gets a set of topologically simultaneous points that fill the outside of the light-cone emanating from the point at which this criterion is being applied (see Figure 4.3). However, because topological simultaneity is not an equivalence relation, the definition of the present must be indexed to each individual spatial position as well as each moment in time. It must be noted then that, as part of

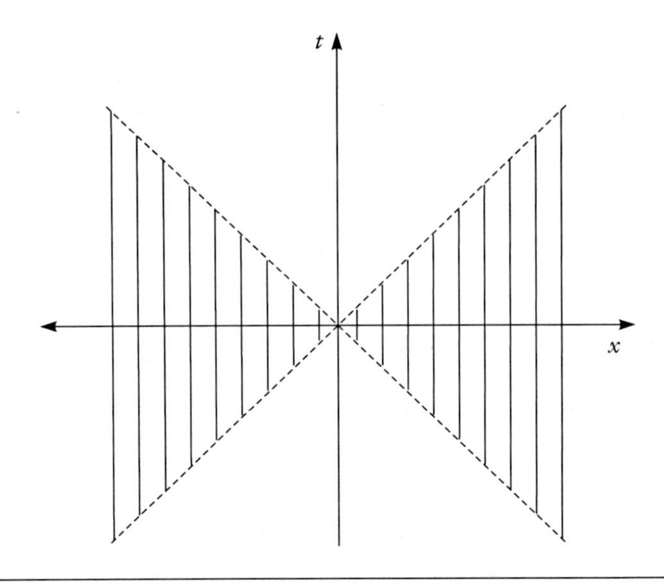

Figure 4.3. The striped area depicts a region of topologically simultaneous points.

Reichenbach's legacy to the discussion of conventions in physical theory, he supplies us with two alternative but related ways to think about simultaneity. The first is the expression that has become synonymous with the conventionalist thesis, and the second his own solution, which was to adopt a new definition of simultaneity, topological simultaneity.

The Conventionality Thesis Refuted?

The conventionality thesis, as introduced by Reichenbach and discussed in the work of Grünbaum, Winnie, and others, is considered by many to have been decisively refuted by David Malament in his 1977 paper, "Causal Theories of Time and the Conventionality of Simultaneity."[11] In this paper, as the title suggests, Malament presented a proof that the Einstein standard of simultaneity ($\varepsilon = ½$ in Reichenbach's notation) could be uniquely defined in terms of the causal structure of Minkowski spacetime. He defined simultaneity, Sim_O, as an equivalence class under the group of "*O* causal automorphisms," for an inertial worldline *O*. In perspectival invariantist terms this is equivalent to considering a set of simultaneous points in spacetime as an invariant structure under a group of symmetry transformations, or alignment criterion. Since this group was to be defined in reference to an inertial worldline, roughly speaking the path through spacetime of a body at constant velocity, he considered the symmetries he felt preserved the necessary causal structure of spacetime with reference to that worldline, a consideration with which Reichenbach would likely have been sympathetic. Malament's analysis leads to the selection of two possible choices for simultaneity criteria, the light-cone (equivalent to $\varepsilon = 1$ and 0) and Minkowski orthogonality ($\varepsilon = \frac{1}{2}$). He then rejects the null cone ($\varepsilon = 1$ and 0) as a trivial simultaneity relation since it implies that all points are simultaneous to each other. This is a consequence of the requirement that Sim_O be an equivalence relation, whereby it must be transitive. Thus, one is left with only one acceptable simultaneity criterion in Minkowski orthogonality, the Einstein convention. According to Malament's argument, one is therefore left with no ambiguity over simultaneity in a given frame of reference (associated with inertial worldline *O*), and therefore no room for conventional choice in the matter.

Malament's relatively short and elegant proof has generally been taken to demonstrate that simultaneity in special relativity is not conventional after all. In addition, it has been held to confirm the predilection many have had toward Einstein's standard of synchrony because of its simplicity, i.e., simplicity as the only choice of ε that may be employed as a convention by observers located at position *A* as well as at position *B*.[12] Furthermore, Malament's proof, by speaking in terms of automorphisms, is making explicit a sense in which this standard of synchrony is objectively definable.

Malament's proof is, in fact, another example of how invariance under a group of transformations has been used to argue for a kind of objectivity. Because the automorphisms under consideration here admit of interpretation as changes in perspective and are heuristically generalizable, one may consider this case from a perspectival invariantist perspective. In introducing perspectival invariantism, it was suggested that debates over the objectivity of a representation play themselves out partly through the choice of what to consider as the relevant group of automorphisms (or invariance criterion). This very tension is evident in recent responses to the details of Malament's proof.[13] This has been attempted by Sarkar and Stachel, who suggest that the group of "*O* causal automorphisms" that he considers should not include the temporal reflection symmetry transformation. This, they argue, is due to the fact that a distinction can be made between past and future light-cones in terms of their causal connection and that these should be considered as two distinct simultaneity criteria, each of them equivalence relations. If this is the case, then one is left with three causally definable choices for standards of synchrony, and there remains a conventional choice between them.

Although Sarkar and Stachel do not make use of Reichenbach's notation, one might have guessed this result from what has already been said, since taking the complete (combined past and future) light-cone as a simultaneity criterion is equivalent to the choice of $\varepsilon = 1$ *and* 0. If each value of ε implies a distinct simultaneity criterion as has been claimed by the conventionalists, then this is clearly the combination of two such criteria. Reichenbach's topological simultaneity has already demonstrated that such a combination may be chosen (in his case a combination of all possible criteria), but at the cost of transitivity, and therefore equivalence.

Sarkar and Stachel conclude their critique of Malament by pointing out that in order to retain these three causally definable criteria, in Reichenbach's terminology $\varepsilon = ½, 1$, and 0, as invariants, then the group of causal automorphisms must exclude time-reversal symmetry. This should be apparent, as the combined forward and past light-cone is mapped into itself by time reversal, whereas $\varepsilon = 1$ and $\varepsilon = 0$ are not. See Figure 4.4.

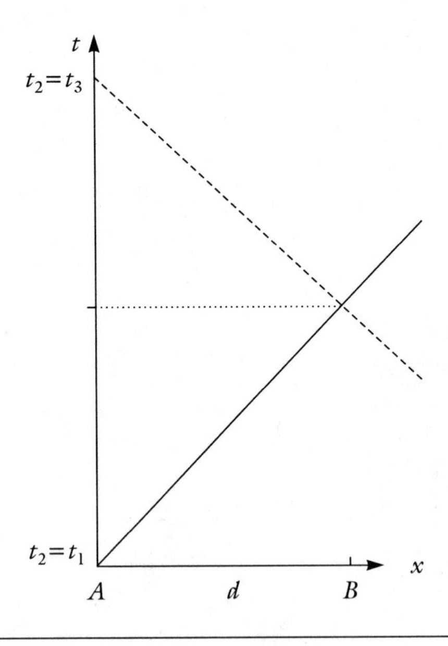

Figure 4.4. The combined simultaneity criteria $\varepsilon = 1$ (so that $t_2 = t_3$) and $\varepsilon = 0$ (so that $t_2 = t_1$) are not time-reversal invariant; the light-cone itself is.

This invariance-based argument illustrates some of the features of symmetry and invariance previously discussed. Among these is the observation that the number of transformations in a given group of automorphisms gets smaller as the number of invariants increases. Removing time-reversal from the group of automorphisms results in three invariant causally defined simultaneity criteria instead of one. The implication for the conventionality thesis is that one may not have a uniquely definable simultaneity criterion after all; now one must choose between one of three options, or, which leads to the same conclusion, choose which group of symmetry transformations to consider as O causal automorphisms. On the basis of these criticisms of Malament, one may argue that conventionality of simultaneity still exists in the choice among the three causally definable simultaneity criteria. This, however, presents a greatly diminished range of conventional choice when compared with the standard claims of the conventionality thesis, as initially introduced without explicit consideration of any causal automorphisms.

Whether or not one is inclined to accept these criticisms of the details of Malament's proof, there is a more elegant response to his rejection of the conventionality thesis that actually revives the conventionality thesis in its fullest sense. This response begins by granting that Einstein synchrony has a unique significance *within a given inertial frame.* This approach, which goes back to the work of A. A. Robb, goes on to claim that even if one accepts Malament's proof as it stands, one is left with an infinite number of conventionally acceptable choices for simultaneity criteria.[14] This is a simple consequence of the fact that Malament has couched his whole discussion by defining synchrony in terms of Minkowski orthogonality with reference to an inertial worldline O. If one takes as a reference another inertial worldline O' that intersects with O, then Minkowski orthogonality will pick out a different set of simultaneous spacetime points. See Figure 4.5. One is thus left with the conventional choice of whether to use the line of simultaneity defined by O, O', or any one of any infinite number of inertial worldlines. In this ambiguity, one returns to what is generally described in textbooks as the relativity of simultaneity. On this account, however, one is faced with the observation that the choice between these different standards of synchrony is a conventional one. This simple observation brings back into consideration all of the implications of the conventionality thesis.

There is a sense in which this new expression of the conventionality thesis is qualitatively different from its traditional expression. Traditionally, the conventionality thesis does not involve the discussion of more than

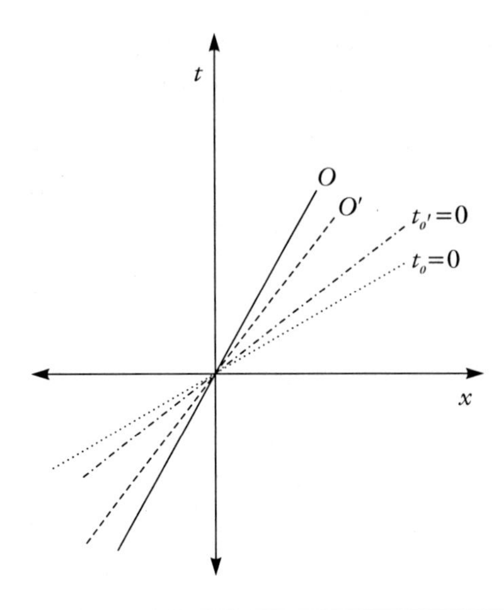

Figure 4.5. Worldlines O and O' are associated with distinct Minkowski orthogonal standards of synchrony.

one inertial worldline or their associated frames of reference. This was instead the domain of the relativity of simultaneity, the canonical textbook approach to simultaneity in special relativity. Sarkar and Stachel's criticism aside, Malament's proof asserts that every frame has a uniquely definable standard of synchrony; this is consistent with the relativity of simultaneity, which expresses the fact that these are each relative to a given frame. Conventionality of simultaneity in this new form thus rests on the familiar consequences of the relativity of simultaneity. In this case, the conventionality thesis and the relativity of simultaneity are equivalent. This sounds potentially contradictory in a context where these have often been juxtaposed as antithetical approaches to understanding synchrony. Even so, it may be shown that the relativity of simultaneity may be understood as equivalent to the conventionality of simultaneity.

Formal Equivalence

Having suggested that the relativity of simultaneity presents one with a conventional choice of sorts, it can be shown that this is in fact formally equivalent to the choice presented by the conventionality of simultaneity.

This might not seem obvious at first. In the case of the former, one begins with standard Einstein synchrony in each inertial frame and must choose which frame to take as a standard across all frames. In the case of the latter, however, one only ever considers a single inertial frame and must choose a simultaneity convention associated with a purported signaling operation. These are two very different scenarios, and in fact this points in the direction of two distinct philosophical concepts of simultaneity, one global and another more local.[15] However, at the level of the formalism of special relativity, there is an equivalence between the two.

This equivalence is manifested in the isomorphism of the structures that represent a simultaneous moment, or time-slice. These structures are distinguished by the hyperplanes of simultaneity that one can construct using either method, the relativity or conventionality of simultaneity. In their most general sense, these hyperplanes, depicted as tilted lines in 1 + 1 dimensional Minkowski spacetime (see Figure 4.6), are simply sets of points between which the equivalence relation of simultaneity is meant to hold.

As a consequence of the relation being an equivalence relation in these cases, hyperplanes of simultaneity have the special property of being partitions of the set of spacetime points. This means that they do not overlap one another and that taken together they fill the whole of space. Another phrase used to capture this partitioning feature of the hyperplanes of simultaneity in relativity is to say that they foliate the spacetime. Because of the way that hyperplanes (the leaves of this foliation) fit together so tightly, a foliation can be characterized by a single leaf, the remainder being simply parallel to the first. It is in terms of this foliation that the relativity of simultaneity and the conventionality of simultaneity can be seen as equivalent.

Starting with the relativity of simultaneity, one may recall that, following Malament, for each inertial frame a single direction can be defined in spacetime without involving assumptions about the one-way speed of light. This is determined by Minkowski orthogonality and delivers, for any inertial frame, what is usually taken as the spatial axis. This then provides a basis for a causally defined foliation of hyperplanes of simultaneity. There is a different foliation, however, for each particular frame. In 1 + 1 dimensional Minkowski spacetime these foliations, or time slices, are depicted as sets of parallel lines. For a resting frame, as viewed from within that frame, the foliation (associated with standard synchrony) is the set of lines parallel to the spatial (horizontal) axis.

However, according to the relativity of simultaneity, all moving frames, as viewed from the rest frame, will appear to have tilted horizontal spatial axes. Each one of these tilted axes defines an infinite set of parallel hyperplanes of simultaneity. So one is left with an infinite number of foliations, characterized by tilted x' axes. These range in slope from $-1/c$ to $1/c$, which is to say that the tilted axes fill up the light-cone. See Figure 4.6. The slope of each of these horizontal axes may be derived from the standard Lorentz transforms by setting $t' = 0$ as follows.

$$t' = \gamma(t - vx/c^2) = 0$$
$$\Rightarrow t = (v/c^2)x$$

So that the slope, m, of the lines of simultaneity in 1 + 1 dimensional Minkowski spacetime is given by:

$$m = (v/c^2) \tag{4.2}$$

where the velocity, v, associated with each inertial frame ranges from $-c$ to c. Finally, one may recall that the choice of which line to accept as defining a foliation of simultaneity hyperplanes is one of convention. This conventional choice involves which frame to take as defining synchrony for all the rest.

Turning back to the conventionality of simultaneity, one begins by

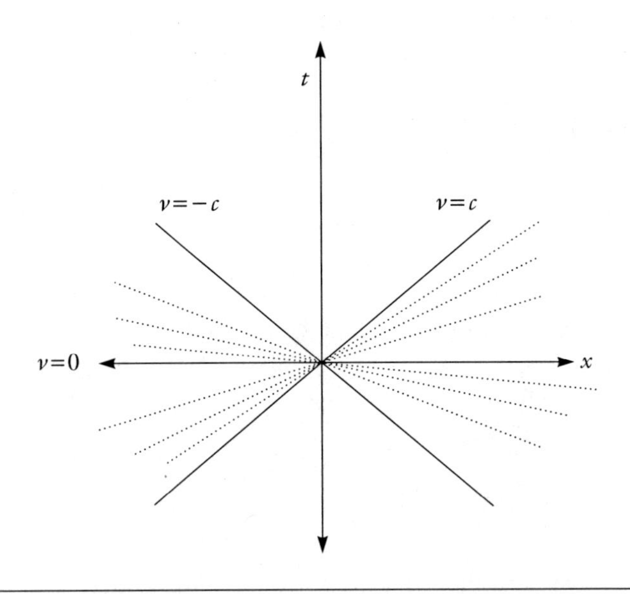

Figure 4.6. Hyperplanes of simultaneity parameterized by velocity.

considering the synchronization procedure used to foliate spacetime. In this case, one only need consider a single frame in which to perform a distant synchrony operation. The assumption that the speed of light is constant in both directions along this path is not made, and the formula may be applied as above in the conventionality thesis (4.1). For a given choice of the parameter ε, simultaneity can be established between spatially distant clocks. If one considers the distance between the clocks to range from 0 to an arbitrarily large distance, then a continuous set of points can be designated as simultaneous.

Thus one may construct all of the same foliations as in the case of the relativity of simultaneity (see Figure 4.7) using different values of ε. The slope, m, of each of these hyperplanes of simultaneity is given by

$$m = 1 - 2\varepsilon \tag{4.3}$$

This, together with the constraint on ε to the interval $[0,1]$ allows one to conclude that the slope ranges from $-1/c$ to $1/c$, as above. Following the same approach as above, each of these hyperplanes, characterized by a value of the parameter ε, can be used to define a foliation of the spacetime. In both cases the lines of simultaneity fill up the light-cone.

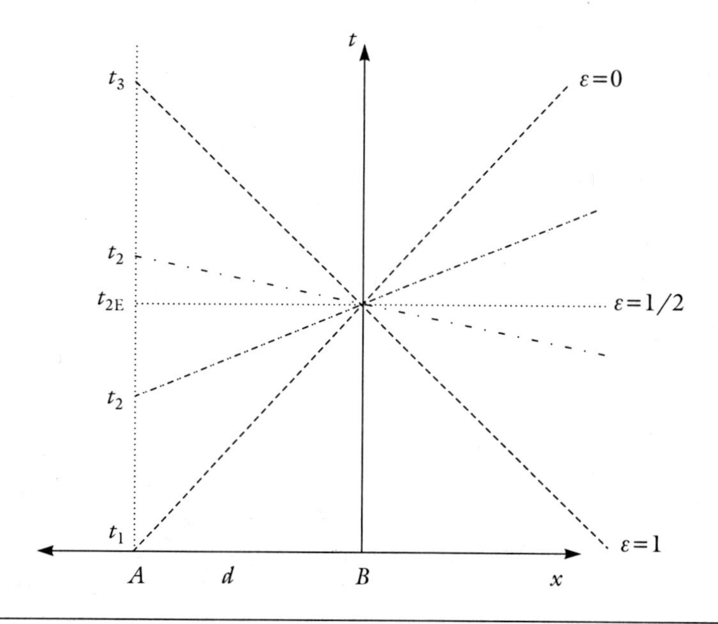

Figure 4.7. Hyperplanes of simultaneity parameterized by ε.

These lines are the same, and constitute the same set of different foliations of spacetime. In the case of the relativity of simultaneity the various foliations are parameterized by the velocity of the frame with which they are associated, whereas in the case of the conventionality of simultaneity they are parameterized by the choice of ε.

The relationship between these two parameters is given by setting slopes equal to one another for a given tilted hyperplane of simultaneity. By combining (4.2) and (4.3), and setting $c = 1$, one finds that:

$$v = (1 - 2\varepsilon), \text{ or equivalently, } \varepsilon = 1/2(1 - v) \tag{4.4}$$

This allows one to link, for instance, the velocity of a moving frame with its implied simultaneity criterion in terms of ε. Furthermore, by substituting values of velocity, v, which range from $-c$ to c, one gets values of ε which range from 0 to 1, as expected.

This should clarify the sense in which the relativity of simultaneity and the conventionality of simultaneity are actually providing alternate methods of accomplishing the same task, that of foliating Minkowski spacetime with hyperplanes of simultaneity. This formal equivalence supports the suggestion already made that, even if one accepts Malament's argument that Einstein synchrony has a special (causally defined) role in a given inertial frame, the ambiguity inherent in the conventional approach to simultaneity is nevertheless undiminished. The question remains, however, of what sort of conventional ambiguity this is.

Which Conventionality Is This?

The ambiguity in choice of simultaneity criterion described above is an example of absolute conventionalism, the claim that there is no factual way to distinguish one from the other. In order to illustrate further what is meant by claiming that this is an example of absolute conventionalism, consider the suggestion that the conventionality thesis (4.1) might be extended. In particular, it might be suggested that since choice of ε is entirely conventional, one might give it any value whatsoever.

This sort of proposal may be acceptable in the trivial sense that one is free to coordinatize spacetime in any way one chooses. However, the use of the Reichenbach notation, with ε ranging only from 0 to 1, does not admit these possibilities because Reichenbach and other proponents of the conventionality of simultaneity were committed to a particular understanding of causation.[16] The limitations imposed by restricting ε to the interval [0,1] reflect this commitment. In fact, unlike Reichenbach, Adolf Grünbaum advocates the conventionality of simultaneity explicitly on the basis of his ontological convictions with respect to the causal structure of events in spacetime.[17] Incidentally, it is the emphasis on the causal definability of simultaneity that made Malament's critique of the conventionality thesis convincing to so many.

One way to illustrate this is to look at the implications for the one-way speed of light in the context of the synchrony procedure. If it is the case that ε is outside [0,1] and the average of outgoing and incoming signal speeds is constant and equal to c, then light must travel at a negative speed on one leg of the journey. A negative speed implies that the signal is moving backward in time, a conclusion that is, to say the least, suspect in light of the basic assumptions of relativity theory.

This objection may be countered by noting that even when ε is within the interval [0,1], there are implications about one-way speeds that seem to violate relativistic assumptions. Taking for example $\varepsilon = 0$, this implies an infinite speed of light in the outgoing direction that might at first seem similarly abhorrent. This impression would be mistaken, however, since a negative speed of light implies that the one-way light signal actually violates deeper intuitions about causality.

The implied propagation of a signal, clearly intended to be a causal influence, backward in time would upset an account that takes the temporal order of local events seriously. Returning to the standard notation in (4.1), an ε value out of the range [0,1] would imply that t_2 is no longer between t_1 and t_3. In addition, the topological notion of simultaneity would become entirely trivial in this situation. That is to say that if ε may be any real number, then every point on the spacetime manifold is topologically simultaneous to every other. Furthermore, since one of the features of a causal chain of events is that its events have a fixed time order, then this would not allow the possibility of strictly causal chains.

This discussion of the possible value of ε is useful to point out that the conventionality of simultaneity does not mean to imply complete freedom of choice. It is precisely the standard hyperplanes of simultaneity, equally well presented in terms of the relativity of simultaneity, which are of interest to the conventionalists. These are parameterized only by requiring ε to be in the range [0,1]. If one wishes to extend this constraint, one could recoordinatize the Minkowski plane by simply switching spatial and temporal axes, but such a choice would clearly be an example of trivial conventionality. Although one can technically pursue this trivial route, the original intent of the proponents of the conventionality of simultaneity was to represent a non-trivial feature of special relativistic kinematics. Thus, while the claim in (4.1) is absolutely conventional, this is not an "anything goes" type of conventionality.

As previously discussed, convention in physical theory comes in different forms. Although they have been considered in various ways by others, here they have been divided into trivial, absolute, and relational conventions. The conventionality of simultaneity exemplifies elements of all three. Trivial or semantic conventionalism is evident, for instance, not only in the obvious (trivial) sense of the semantics of technical language, but in the fact noted above that one must choose a set of coordinates in which to even express the conventionalist claim. In Reichenbach's terms the coordinative definition associated with choosing units of time and space, by association with physical rods and clocks, constitutes what has been called relational conventionalism above.[18] Finally, absolute conventionalism is evident in the claim stemming from the conventionality thesis (4.1) that there is no fact-of-the-matter according to which one may prefer a specific choice of simultaneity criterion.

Simultaneity and Perspectival Invariance

The tradition, briefly sketched in this chapter, of the conventionality of simultaneity has been significant as the paradigmatic example of a conventionalist claim in modern physical theory. It has, moreover, been of some use in helping to clarify some interpretative difficulties within relativity theory and the philosophy of time, especially with regard to the notion of temporal becoming. In addition to its implications for the understanding of time, however, the conventionality thesis also illustrates some of the salient features of the general use of conventions in science.

For these reasons, Salmon is correct when he suggests that "this controversy . . . is an excellent testing-ground for a number of philosophical doctrines concerning conventionality."[19] In fact, it also serves as a good initial illustration of perspectival invariantism. That the case of representing simultaneous events in special relativity admits of such an interpretation should be evident from the fact that the symmetries involved may be easily interpreted as perspectives and that these symmetries have proven to be generalizable and immensely heuristically successful.

However, not only does this serve as an illustration of perspectival invariantism, but it provides a new insight into the philosophical point at issue between proponents of the conventionality of simultaneity on the one hand and the relativity of simultaneity on the other. For the perspectival invariantist, the difference between these views can be understood as a disagreement over what precisely is objective *(Obj_P)* in special relativity.

In addition, the perspectival invariantist approach to this controversy demonstrates that here invariance has at least as much to do with convention as it does with objectivity. Thus, this case stands as evidence in favor of the claim that there need be no fundamental tension between objectivity and convention in scientific representation. Although perspectival invariantism only works in specific circumstances, where it does work it always illustrates this general point about the relationship between objectivity and convention.

To draw out these points in greater detail, one may begin by stepping back to consider what is really at stake in the perennial debate over the conventionality of simultaneity. One way to answer this question is in terms of the objectivity of the simultaneity relation. The conventionalist claims it is absolutely conventional, and the proponent of the relativity of simultaneity that it is objective, in a given frame. Taking a perspectival invariantist approach to this debate, it becomes clear that, in different senses, both are correct. Indeed, perennial debates often emerge where each competing point of view is justifiable on the basis of its own assumptions. Thus the real debate must be over these assumptions. In this case, for the perspectival invariantist, the key assumption is over which group of symmetries is taken as the relevant invariance criterion.

According to the perspectival invariantist, there is an objective sense to the assertion that two points are simultaneous. The idealized model of spacetime provided by special relativity is as an abstract manifold of points, each representing events in space and time. The mathematical model is a concrete structure, usually referred to as an inertial frame. As has been shown, these frames may be parameterized by a velocity, as in the Lorentz transformations. This is effectively Malament's approach, whereby he specifies an inertial frame by the worldline of an object moving at some constant velocity. Moreover, a standard of synchrony is built into each one of these inertial frames through the mathematical relation of Minkowski orthogonality, or the Einstein convention.

In order to answer the question of whether any of these standards of synchrony are objective, the perspectival invariantist must specify the relevant group of symmetries that will serve as an invariance criterion. As previously suggested, the spacetime symmetries of special relativity are represented by the Poincaré group, which includes among its members spatial translations, rotations, and the Lorentz transformations. These symmetries may each be easily interpreted as mappings between idealized spatiotemporal perspectives. What is more, they have proven to be generalizable and heuristically fruitful symmetries. In these circumstances, the perspectival invariantist would be able to claim that features of the spacetime manifold that are invariant under the Poincaré group are objective *(Obj$_P$)*.

Thus one might ask if the simultaneity relation for each inertial frame, given by standard Einstein synchrony defined in that frame, is objective *(Obj$_P$)*. Since Einstein synchrony is represented by a hyperplane of simultaneity that is Minkowski orthogonal to an inertial worldline, the perspectival invariantist must simply check to see if this structure is invariant under the relevant invariance criterion, the Poincaré group. As it happens, this hyperplane is not invariant under transformations from one inertial frame to another; it is not Lorentz invariant. This is apparent from the fact that Minkowski orthogonality with any other inertial worldline defines a different hyperplane.

Yet many feel there is something special about Einstein synchrony that should lend it a kind of objective status. For Malament, the best-known champion of this position, this is causal definability. Perhaps surprisingly, the perspectival invariantist may still be able to agree with this intuition by specifying a different invariance criterion. This may be accomplished by simply removing the Lorentz transformations. The resulting set of symmetries still qualifies as a group, since it is still closed under multiplication, includes an inverse for each member, and possesses an identity element. Under this group, the foliations defined by Einstein synchrony are clearly invariant, and are therefore objective *(Obj$_P$)*.

In fact, on a more careful reading this is exactly the sense in which Malament attempts to demonstrate there is something special about standard synchrony. Malament concludes that given a worldline, there is a uniquely definable standard of synchrony. Since this standard depends on the specification of a worldline, Malament and those who follow him on this point are in reality arguing only that standard synchrony is objective in a given frame. By pinning down this notion of standard of simultaneity to a given frame, one is in effect removing Lorentz transformations from consideration. This is reminiscent of the case already discussed of the objectivity of a vertical direction. In that case, it was pointed out that it is impossible to spatially translate the vertical direction at position *(x,y,z)*. It is similarly impossible to Lorentz boost the hyperplane of simultaneity in frame F, since the resulting structure will be in some boosted frame, F′. Thus, this approach is equivalent to removing Lorentz transformations from the invariance criterion. This done, the perspectival invariantist may agree with Malament that there is something objective *(Obj*$_P$*)* about Einstein synchrony, but not in reference to the full symmetry group of special relativity.

However, the perspectival invariantist is likely to prefer to use the Poincaré group as an invariance criterion; the Lorentz transformations are, after all, some of the most generalizable and heuristically powerful symmetry transformations in modern physics. If the perspectival invariantist makes this choice, then he or she may say that standard synchrony is objective in a limited sense, but in the full sense of special relativity, it is ultimately conventional.

Thus, the debate over the conventionality of simultaneity illustrates a case in which perspectival invariantism does apply. In fact, adopting this approach sheds new light on the key assumptions at issue here, namely which symmetries to use as an invariance criterion for objectivity *(Obj*$_P$*)*. In addition, since the objectivity of the synchrony relation depends on this choice of invariance criterion, one can see that for the perspectival invariantist, invariance has at least as much to do with convention as it does with objectivity. In the two remaining case studies, many of these issues raised by the debate over the conventionality of simultaneity will resurface in ways that support this general perspective on the relationship between objectivity invariance and convention.

5

Objectivity in the Twin Paradox

In the previous chapter, the attempt to represent simultaneous events in special relativity was considered in some detail. A second case, which may also be used to illustrate a perspectival invariantist approach, is based on the attempt to represent the elapsed time on a given path through spacetime. The theory of special relativity maintains that there is a mathematical model of this idealized notion; the representational relation between model and idealization has sometimes been called the "clock hypothesis."[1] According to this proposal, the elapsed time, t_e, on an ideal clock, as it moves over a certain path in space, is represented by a calculated quantity, τ, called "proper time."

In Chapter 4, it was suggested that in order for a representation to be objective, the elements of its formal chain of media (including an original and a model) must remain invariant under a given group of symmetry transformations, termed an invariance criterion. In the case of the clock hypothesis, the invariance criterion is the standard invariance group of Einstein's special theory of relativity, the Poincaré group. Thus all representations that establish the representational relation expressed by the clock hypothesis may be considered objective *(Obj$_P$)* with respect to this invariance criterion.

The most famous example used to establish the relation between elapsed time and proper time is that of the twin paradox. By comparing various accounts of the twin paradox, one may note the ways in which various ambiguities are exploited. One may also note how these ambiguities may be resolved by conventional choices; these choices include especially the absolute conventionality with respect to the notion of simultaneity discussed in the previous chapter.

In order to further illustrate the implications of perspectival invariantism, the twin paradox will be considered in detail and a generalized method of comparing different accounts of it will be proposed. In doing so it will be demonstrated that each account shares the invariant representational relationship between elapsed time, t_e, on a moving clock and proper time, τ. Furthermore, it will be shown how the various accounts of the twin paradox in the literature exploit the ambiguity provided by different standards of synchrony allowed by the conventionality of simultaneity. Finally, it will be concluded that the constraint imposed by the objectivity (Obj_P) of the representational relation between t_e and τ delimits the scope of conventional choice available to the physicist in the telling of the story of the twin paradox.

An Introduction to the Twin Paradox

This case study is focused on the twin paradox of special relativity as an example of a scientific representation in modern physics. The starting point will be a brief sketch of the minimal standard version of the twin paradox, including some of the calculations involved and a discussion of the primary representational relation it attempts to establish as indicated by the clock hypothesis.

Origins and a Minimal Standard Version

Einstein's theory of special relativity leads to a conception of space and time that appears counterintuitive from a pre-relativistic perspective in a number of ways. None of these is more fundamental than the path dependence of proper time calculated along a given worldline, a series of connected points in spacetime. In addition, no illustration of this point is better known within and outside of the physics literature than the twin paradox. This scenario was first suggested, though not specifically called the twin paradox, in 1911 by the French physicist Paul Langevin in a paper entitled, "L'évolution de l'espace et du temps."[2] Langevin, writing during the few years following Einstein's introduction of special relativity in 1905, was attempting to come to terms with the full implications of relativistic kinematics. The response of the relevant physics community to Einstein's work was far from unified at this time, especially when compared with the extent to which special relativity and its problems later became part of the canon of physics education.

The first discussion of what came to be called the twin paradox in Langevin's work also followed shortly after the significant 1908 paper of the mathematician H. Minkowski, who managed to make explicit the unexpectedly intimate connection between space and time expressed in the spacetime of special relativity.[3] The formalism by which he achieved this, the construction of a spacetime manifold, became the standard means of presenting conclusions about relativistic kinematics. It is also the case that the flat spacetime of special relativity, which bears Minkowski's name, provided a foundation for consideration of curved spacetimes in Einstein's general relativity. Although Langevin does not make explicit use of Minkowski's approach to relativistic spacetime in his paper on the twins, this discussion of the twin paradox will be presented, consistent with current convention, in terms of flat Minkowski spacetime.

In fact, Langevin's 1911 presentation of the twin paradox was entirely free from explicit mathematical formalism. Instead, he recounts the story of an astronaut who journeys to a distant star at a speed near that of light and returns home. Upon his return, he discovers that although he has been traveling for two years according to his shipboard chronometer, two centuries seem to have elapsed back on Earth. This dramatic narrative is replete with descriptive detail reminiscent of late-nineteenth-century science fiction. As a result, the tone of the 1911 paper is quite unlike that of Einstein's own presentation of special relativity in 1905. Langevin intended his work to be an *"exemple concret"* of Einstein's theory.[4]

To this basic plot has often been added the idea that the space traveler was one of two twins, the other of whom stayed on Earth. Thus the traveling twin returns home to discover that the earthbound twin has aged much more quickly and that the ages of the two twins are no longer the same. The story of the twins and their differential aging can be depicted geometrically on Minkowski spacetime. This is typically accomplished by translating Langevin's account onto a flat spacetime with one dimension each for time and space (often referred to as $1 + 1$ dimensional spacetime). This is usually depicted with the time axis, t, running vertically up the page and the space axis, x, running horizontally from left to right. Each twin is associated with a worldline, the first simply running vertically along the time axis, corresponding to the twin who stays on Earth. The second worldline, corresponding to the traveling twin, shares its two endpoints with the first, depicting the twin's departure and return. Between these endpoints, however, it is not entirely clear how best to translate Langevin's story into the formalism of special relativity. There are numerous ways in which this has been and can be accomplished, some of which will be considered in detail. However, traditionally the path of this second worldline frames a regular triangular shape, the basic features of which reflect the notion that the traveling twin moves at some constant velocity away from Earth and then reverses direction and returns at the same speed. See Figure 5.1.

Having translated Langevin's story of the twins into two worldlines on Minkowski spacetime, the formalism of special relativity allows one to calculate the proper time along each of these paths. This is most simply expressed by pointing out that the proper time, τ, along any worldline can be calculated as the path integral through spacetime of the infinitesimal, $d\tau$, where

$$d\tau = \gamma^{-1} dt \tag{5.1}$$

The term γ is the same as that in the standard Lorentz transformations and is given by

$$\gamma = (1 - v^2 / c^2)^{-\frac{1}{2}} \tag{5.2}$$

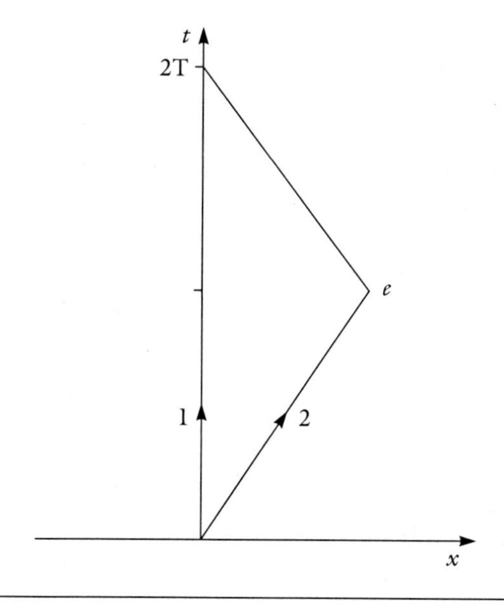

Figure 5.1. The minimal standard version of the twin paradox in (1 + 1 dimensional) Minkowski spacetime. The labeled arrows indicate the paths of the two twins.

Since the velocity, v, must be less than the speed of light, c, then γ is a nondimensional parameter greater than or equal to 1; alternatively, its multiplicative inverse, γ^{-1}, which appears in (5.1), is less than or equal to 1. In the limit as v approaches the speed of light, then γ^{-1} approaches zero:

$$0 \leq \gamma^{-1} \leq 1. \tag{5.3}$$

According to expression (5.1) above, the parameter, γ^{-1}, then generally associates intervals of coordinate time, t, along a worldline with smaller intervals of proper time. The exception is in the trivial case in which the velocity, v, is zero, which implies a stationary worldline; in this case proper and coordinate time intervals would be equal to one another.

In this specific case one could conclude, from the clock hypothesis, that coordinate time, t, is the same as the elapsed time, t_e, on the earthbound twin's clock. In general, however, proper time and coordinate time differ by the factor γ^{-1}, as above in equation (5.1), so that proper time is in a sense a combination of spatial and temporal intervals. By combining (5.1) and (5.2) one may see that

$$d\tau = \left(dt^2 - v^2 dt^2 / c^2\right)^{\frac{1}{2}} = \left(dt^2 - dx^2\right)^{\frac{1}{2}}. \tag{5.4}$$

An interval of proper time is equal to the square root of the difference of the squares of the intervals of time and space. This elementary feature of special relativity is one way of illustrating its so-called "meshing" of the concepts of space and time into that of spacetime.

Applying this formalism to the story of the twins, one can calculate the proper time along the two simple worldlines described above. By employing the following simple integral expression for proper time along a worldline with beginning point b and ending point e,

$$\tau = \int_b^e \gamma^{-1} dt, \tag{5.5}$$

one can arrive at two proper times, τ_1 and τ_2, associated with earthbound and traveling twins, respectively. The worldline associated with the former is precisely the trivial case mentioned above in which intervals of coordinate and proper time are the same length. The worldline associated with the latter leads to a different proper time value:

$$\tau_1 = \int_b^e dt = 2T > \tau_2 = \int_b^e \gamma^{-1} dt = \gamma^{-1} \int_b^e dt = \gamma^{-1} \cdot 2T. \tag{5.6}$$

The proper time of the first worldline, τ_1, associated with the earthbound twin, is indeed greater than that calculated along the second, τ_2, associated with the traveler. These calculations, taken together with the simple diagram, Figure 5.1, of the twins' paths through spacetime, form a minimal standard version of the twin paradox. This may be used as a basis of comparison for the numerous different accounts of the twins that have been given. Before moving on to this, however, one must clarify the primary representational relation that all accounts of the twin paradox seek to establish.

The Clock Hypothesis as a Representational Relation

The twin paradox is considered here as an illustration of scientific representation and to this end one can specify the primary representational relation involved as that between the elapsed time measured on a clock, t_e, and the proper time, τ, measured along the path of that clock through spacetime. Because both t_e and τ are simple structures, intervals on the real number line, they are often referred to simply as quantities. For the purposes of this case study one may think of intervals like these fundamentally as representational structures. Furthermore, it will be shown how the clock hypothesis formalizes the representational relation between them. In the process, another feature of representation previously discussed will be made explicit; namely, the formal chain of media, which in the case of the twin paradox includes actual clocks, ideal clocks, and proper time calculations.

Accepting the calculations in the previous section, one may additionally assume that proper time represents the elapsed time on a clock traveling along any given worldline. Further, one may assert that each twin's biological age is the same as that measured by each twin's clock. These steps lead to the conclusion that the twins age differently. Depending on the specific velocities and the distance of the voyage, one can, for example, reproduce Langevin's original scenario, resulting in a particularly large gap of 198 years between elapsed time on Earth and on the astronaut's ship, respectively. By linking biological age to clocks, a statement about the calculated quantity proper time may be taken to represent age in the normal conversational sense of the term.

As previously noted, identifying proper time and clock measurements has been called the "clock hypothesis." Thinking in this way about clocks traveling along worldlines is characteristic of the discussion of these issues in relativity; in keeping with this approach, discussions of relativity have sometimes addressed this scenario as the "clock paradox" instead of the twin paradox.[5] Either explicitly or implicitly, the simplification of the story to one about clocks instead of twins is, in L. Marder's words, simply an attempt to "avoid . . . the [biological] issue of whether a traveler's ageing is in accord with the standard clock" that travels the same worldline.[6] The implication in the twin paradox that a person, as a physical system functioning periodically, is somehow a clock, and therefore also a structure in the chain of media, raises interesting issues in itself. Accepting for the sake of this discussion that humans, perhaps a pair of real twins, are physical systems that register an elapsed time, one may attempt a more precise treatment of the clock hypothesis.

In thinking of each twin as aging in agreement with a standard clock, it should be apparent that the clock hypothesis may be taken on at least two different levels. As stated above, the clock hypothesis maintains that elapsed time, t_e, on a moving clock is represented by τ, the proper time. Thus one may claim that clocks measure proper time. This is to say that intervals of elapsed time on a moving clock, one traveling along a given worldline, are of the same magnitude as intervals of proper time calculated along that path. This may, in fact, be broken down into two separate claims by considering the clock in question as either an actual physical system or as an idealized clock in the sense in which Einstein introduces them in special relativity.

One might refer to these two assumptions separately as clock hypotheses 1 and 2. Clock hypothesis 1 may be understood as relating a real clock, a physical system (human or other) that periodically returns to its initial state, to proper time. This is how it is usually understood, and from it might follow a discussion about what sorts of clocks may best be able to realize this possibility physically. The Hafele-Keating experiment, which involved two atomic clocks carried around Earth in opposite directions, was a (successful) attempt to do just this.[7] Although the clock hypothesis is usually taken in this sense, one may fruitfully distinguish this from a second alternative, clock hypothesis 2.

Clock hypothesis 2 asserts that there is a theoretical structure, an ideal clock that is associated with proper time from within the theoretical formalism of special relativity. Thus the chain of media between the elapsed time on an original physical clock and a proper time calculation also includes an interval on an ideal clock. The representational relation established by clock hypothesis 2 is between the ideal elapsed time and proper time, while clock hypothesis 1 adds that there is a link between physical and ideal clock measurements; hence the sense in which the two hypotheses mesh with one another.

Because of the closeness of this interrelation, the distinction between these two quite different types of claim is not generally made explicit. However, returning to Einstein's and Langevin's 1905 and 1911 papers, the distinction is at least made implicitly. As defined above, clock hypothesis 2 should be familiar as an extension of Einstein's treatment of clocks in the 1905 paper. In this paper, Einstein populates space and time with ideal structures, which he calls "rods" and "clocks," that perform the role of establishing ideal spatial and temporal intervals, respectively.[8] Meanwhile, accepting that clock hypothesis 1 is implicit in Langevin's 1911 narrative, then there is also a representational relationship between the ideal rods and clocks and actual rods and clocks. In fact, both actual clocks and ideal clocks are structures that form part of the formal chain of media—a chain that links actual clocks to ideal clocks to proper time calculations.

This picture, and the distinction between clock hypotheses 1 and 2, helps clarify what might otherwise appear as an inconsistency in the treatment of clocks in special relativity. The apparent inconsistency arises because clocks are treated variously as actual physical structures and as mathematical representational structures in different circumstances within relativity theory. In what sense then are rods and clocks taken as actual structures in special relativity? According to his famous 1905 paper, "On the Electrodynamics of Moving Bodies," Einstein begins his discussion of transformations between reference frames by taking "two systems of coordinates, each of three rigid material lines, perpendicular to one another and issuing from a point," and continues, "Let each system be provided with a rigid measuring rod and a number of clocks . . . in all respects alike."[9] The ultimate product is a lattice of points laid out in space using a measuring rod. Each point in three-dimensional space is then associated with a clock synchronized by a process according to Einstein's standard synchrony criterion (in terms of Reichenbach's notation, $\varepsilon = 1/2$; see Chapter 4). The use of rods and clocks in these sorts of synchronization procedures is often taken to reflect the operational nature of Einstein's theory. In this sense, the construction of coordinates with a clock at every point of a lattice formed with rigid rods can be understood as a literal surveying of space and time; rods and clocks are objects, and synchronization is a real procedure within relativistic spacetime.

This way of using rods and clocks to construct coordinates is close to what Reichenbach refers to as establishing a coordinative definition, as discussed in the previous chapter.[10] Illustrating this concept, Reichenbach presents us with the suggestion that a rod may be used to define a unit of spatial distance. Central to his definition is the requirement that the rod be a physical object to which a theoretical structure, in this case intervals of distance, can be associated by convention. As he puts it, "The characteristic feature of this method is the co-ordination of a concept to a physical object."[11] Picking up on this tradition, Lawrence Sklar, for example, explicitly uses Reichenbach's idea of a coordinative definition in his discussion of establishing simultaneity in special relativity.[12] It would appear that, in some ways at least, the rods, clocks, and operations of relativity theory are intended to be actual.

Having noted some ways in which the rods and clocks of special relativity are treated as actual, one may also recall the simple point that special relativity is not presented by Einstein as dealing with dynamics (that is to say with forces) at all, so that its clocks and rods must in some sense also be idealizations. This raises the distinction between kinematics and dynamics; the consideration of masses and associated forces and momenta in relativity theory is traditionally understood as an issue of dynamics.[13] Kinematics on the other hand is usually understood to relate to spacetime representations that do not account for forces and momenta. Dynamical considerations ought, however, to conform with kinematics, in the sense that the dynamical description of moving rods and clocks should be consonant with the kinematical contractions and dilations famously associated with special relativity. Here it is enough to note that the rods and clocks of special relativistic kinematics are not intended to be dynamical objects, i.e., objects that exert and are affected by forces. This seems to suggest that as representational structures they are not physical objects after all, but idealizations.

This is not simply calling attention to the fact that actual clocks invariably fall short of the ideal due to our limited watch-making expertise. It instead makes a stronger claim about the physical systems commonly called clocks. One may recall that a clock is in principle a device for marking out intervals of time at a given fixed position. Actual physical clocks, however, must be conceived, even ideally, as more complicated physical mechanisms. A clock's mechanism must be periodic in that it returns to its initial state after regular intervals, as in some oscillation.[14] Thus clocks as periodic physical systems must be, even in principle, dynamical objects. If this is the case, then there does appear to be a serious tension in the way that clocks in particular are treated within special relativity.[15]

In the twin paradox, clocks are likewise treated as at once actual objects and idealizations. In fact, this is a distinction closely parallel to that between clock hypotheses 1 and 2. One can resolve this potentially problematic tension by pointing out that ideal and physical clocks both play a part in the twin paradox. The function of all these clocks within this context is a part of the collection of structures that make up the formal chain of media. This chain includes both *actual* and *ideal* rods and clocks. In this way one may speak both of actual clocks and ideal clocks without contradiction. Thus one may take the clock hypotheses, both 1 and 2, to establish a representational relation between actual clocks and proper time, as well as a formal chain of media that include idealized clocks. This is the primary representational relation underlying the various accounts of the twin paradox under consideration.

The Invariance Criterion

From a perspectival invariantist point of view, in addition to a representational relation, an invariance criterion must also be specified. This invariance criterion, in the case of the twin paradox, is the usual invariance group of special relativity, the Poincaré group, i.e., the inhomogeneous Lorentz group. If the structures that comprise the formal chain of media are invariant under this group of transformations, then one may claim that a given account of the twin paradox is objective *(Obj_P)*.

Turning to the Poincaré group as an invariance criterion, the features of the members of the formal chain of media must be invariant under this group of symmetry transformations. As previously suggested, this chain includes especially the elapsed time on a moving clock, t_c, and the proper time, τ, calculated along that clock's path. According to the requirement above, these must both be invariant under all of the transformations that make up the Poincaré group, which includes spatial translations, rotations, and Lorentz transformations. The invariance of structures under spatial rotations and translations has already been discussed. In addition, the Lorentz transformations of special relativity, below, relate the spacetime coordinates of an event in one inertial frame to its coordinates in another, moving with constant velocity v with respect to the first. The transformations may be expressed in (1 + 1 dimensional) spacetime as,

$$
\begin{aligned}
x' &= \gamma(x - vt) \\
x &= \gamma(x' + vt') \\
t' &= \gamma(t - vx / c^2) \\
t &= \gamma(t' + vx' / c^2)
\end{aligned}
\tag{5.7}
$$

where the unprimed terms are spatial, x, and temporal, t, coordinates in the initial frame and the primed terms are spacetime coordinates in the moving frame. The term γ is the same as that discussed already in the calculation of proper time (5.2). Lorentz transformations link different possible sets of inertial coordinates in Minkowski spacetime. Therefore a structure is invariant under a Lorentz transformation if it remains the same in any one of these frames of reference. Intervals of proper time are famously the primary invariants of the Lorentz transformations and the clock hypothesis functions to link this interval to a movable clock. Thus, as elements of the formal chain of media, both elapsed time on a clock and proper time are invariant under the group of transformations (the Poincaré group) that forms the relevant alignment criterion.

One might wonder what structures within the story of the twins fail to be aligned in this same sense; i.e., those that fail to be invariant under the Poincaré group. The most obvious examples of this are intervals of space and time. It has been shown that these have been associated with the rigid rods and clocks used to set up inertial coordinates in the first place. As previously suggested, proper time is an interval that is in some sense a mixture of spatial and temporal intervals. However, whereas proper time is an invariant of the Poincaré group, pure spatial and temporal intervals are not.

Another way of expressing this is in terms of the famous relativistic effects of length contraction and time dilation. These consequences of special relativistic kinematics are almost always presented as distinctive features of the theory. A length, often visualized as a rigid rod, is defined as a spatial interval at a given moment in time. A time interval is, on the other hand, traditionally associated with an array of stationary clocks in a given frame of reference. Length contraction is the phenomenon that a rod (interval of space) viewed from another frame is shorter than in its own rest frame, and time dilation the phenomenon that intervals of time on a clock appear to last longer when viewed from a frame other than its rest frame.

In order to illustrate this, one may note, following the Lorentz transformation for the position above, that a spatial interval $[0,b']$ subtended by a rod in a moving frame transforms to the smaller interval $[0,b]$ in the resting frame. If, as is implied, the left end of the rod is considered as the position $x = 0$, then following (5.7), its length in the rest frame may be given by

$$b = \gamma(b' + vt') = \gamma b' \quad (5.8)$$

where t' is zero because the moving rod is defined as an interval in space at a constant time. It is also worth noting that this calculation depends on a clearly defined notion of simultaneity. That b is greater than b' (by a factor of γ) is the essence of the length contraction, but for the purposes of this argument the important conclusion is that intervals of space are not invariant under Lorentz transformations.

Similarly, one may show that the time interval measured by a clock in a resting frame, $[0,t_a]$, transforms to the interval $[0,t_a']$ in a moving frame. See Figure 5.2. The changes of scale are given once more by the Lorentz transformations (5.7) so that

$$t_a' = \gamma\left(t - vx/c^2\right) = \gamma t \quad (5.9)$$

Therefore the time interval, $[0,t_a']$, in a moving frame is larger than the interval $[0,t_a]$ in a resting frame by a factor of γ. This demonstrates that neither pure spatial nor temporal intervals are invariant under the Poincaré

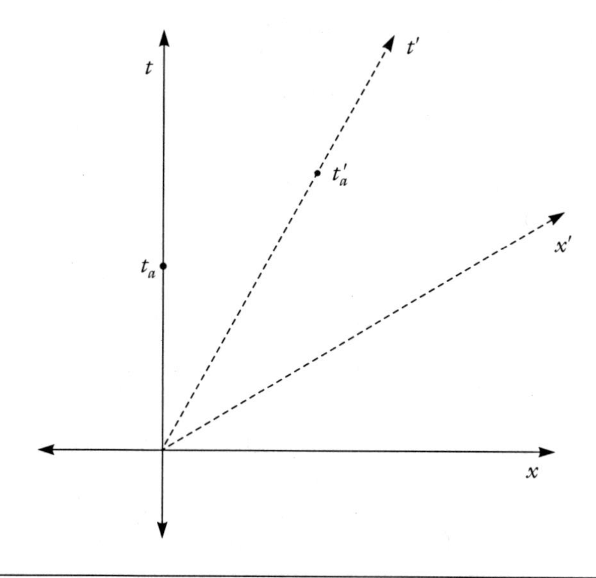

Figure 5.2. Time dilation, the noninvariance of pure temporal intervals in special relativity.

group precisely because they fail to remain invariant under Lorentz transformations.

It may now be pointed out that this is what one should have expected in the first place. These spatial and temporal intervals, in the form of rods and clocks, are the basis on which inertial coordinate systems are constructed, and Lorentz transformations are designed to map one possible set of inertial coordinates, a reference frame, to another. Thus one should expect that a coordinate system, and the intervals of space and time marked out along its axes, is from the outset a structure that changes when acted on by a Lorentz transformation. This is not to say, however, that coordinate systems are somehow superfluous. Without an inertial set of coordinates, it would not be possible to make a calculation of the invariant quantity proper time as in (5.2) above. This does suggest that specific sets of coordinates (built up of spatial and temporal intervals) may not be part of an invariant chain of media for the representation of elapsed time on a moving clock, as occurs in the twin paradox. From a perspectival invariantist point of view, proper time is an objective *(Obj_P)* quantity precisely because it is invariant under the relevant group of automorphisms for special relativity.

This situation illustrates once more the inverse relationship previously discussed between invariants and symmetries. All three structures, spatial intervals, temporal intervals, and proper time are invariant under a smaller group including just spatial translations and rotations. Under the larger Poincaré group, which includes the Lorentz transformations, only proper time intervals remain invariant.

Where Is the Paradox?

Before moving on to look at different accounts of the twin paradox, it is instructive to consider a question that has often been asked about the twins. Given the unproblematic, if counterintuitive, calculation of the proper times along the twins' paths, one might ask whether the twin paradox is actually a paradox at all. Or simply, what's so paradoxical about that? In response, one may take more than one approach to defining a paradox, some more restrictive than others. A paradox may be considered to be a situation that contains an apparent contradiction, or alternatively one in which a logical or even *a priori* contradiction actually holds. Aside from a very small minority of commentators on the story of the twins, it is generally agreed that the twins end up with different ages as a consequence of relativistic considerations.[16] Because of this fact many have been tempted to conclude that in fact no paradox exists since there is no actual contradiction involved.

One might reply that there is still an apparent contradiction in the twins' histories and that the twin paradox might in this sense be a paradox after all. Following this approach, one needs to understand what might constitute an apparent contradiction. There is an apparent contradiction in the scenario of the twins that is a consequence of one of the relativistic effects just considered, time dilation. This effect suggests that time intervals in a moving frame appear longer when transformed into a stationary frame. Since time intervals are measured by the periodic motion of a clock, the slogan "moving clocks run slow" is often applied to describe the situation. In the minimal standard version of the twin paradox discussed up to now, both twins are constantly in motion with respect to one another between the time of the traveling twin's departure and return. Thus, according to the time dilation slogan, each twin might say that the other's moving clock runs slower than his or her own. Yet, clearly this is only the case for one twin, the earthbound twin, at the end of the day. Hence there is an apparent contradiction and the existence of a paradox in the less restrictive sense of the word.

Something like this is surely what has been intended in the countless discussions of the twin paradox in the literature. However, according to a perspectival invariantist account, the twin paradox fails to be a paradox even in the limited sense of being an apparent paradox. This is due to the fact that one may restrict, on the basis of invariance, the representational structures used in objective *(Obj_P)* representation. Recalling that time dilation is based on the failure of pure time intervals (in inertial coordinate systems) to be invariant under the Poincaré group, one may exclude them from the formal chain of media used in objective *(Obj_P)* representations of the twin paradox. This done, the apparent contradiction due to time dilation may be similarly excluded from our account of the twin paradox so that it is not even a paradox in the apparent sense. Thus it is only proper time and moving clocks that provide an objective *(Obj_P)* basis for accounts of the twin paradox.

Explaining the Twins

Since Langevin first told his tale of a space traveler in 1911, there have been many retellings of the twin paradox, either as part of standard textbook introductions to special relativity or within the broader academic literature. Although many textbook accounts are primarily intended as a way of illustrating how to perform special relativistic calculations, all of these accounts are to some degree attempting to provide an explanation for why the twins end up with different ages.

What is perhaps most surprising for such a well-worn problem as the twin paradox is that many of these different accounts are plausible but also seem to contradict one another in important ways. Almost all commentators agree that the twins have experienced different amounts of elapsed time upon the return of the traveling twin, but they part ways considerably in giving an explanation for why this is the case. This situation has led Henri Arzelies to remark that within this debate "the same arguments are always advanced and the same replies given."[17] One may respond to this situation by demonstrating that these numerous accounts of the twin paradox may be treated under a single scheme.[18] Taking this approach, these accounts all appear as special cases of a more general account. It will be helpful to go through some of the arguments leading up to this conclusion.

One may, however, take this discussion even further. It will be argued that the debate over the twins may be most clearly understood as having to do with which explanation is the best explanation for the twins' differential aging, and furthermore that restricting the discussion to objective *(Obj_P)* representations leads to better explanations. This is due to the claim previously made that perspectival invariantism helps one choose better representations. On the one hand, almost all of the usual retellings of the twin paradox fail to be objective *(Obj_P)*. On the other hand, the minimal standard version of the paradox presented above succeeds in being objective because the features of its models are invariant under the Poincaré group. It will be shown that this is the only account that is fully consistent with the generalized scheme presented. Thus one may conclude that representations based only on the representational relation between elapsed time on a clock and proper time provide the best explanation for the twins' aging.

Asymmetries

Starting with Langevin, the basic approach to explaining the different ages of the twins has traditionally been to note asymmetries between the experiences of the twins. That is, how might each twin experience life differently from the other? This approach is not adequate. As it turns out, the *only* asymmetry that remains in an objective *(Obj$_P$)* representation of the twin paradox is the differential aging itself. This leaves one in a bad position indeed if one intends to explain the different ages with reference to the asymmetry that they have different ages! Nevertheless, the search for asymmetries to account for the differential aging has set the tone for the subsequent debate, which has been so conspicuously inconclusive in the literature. In order to get a feel for this debate it is useful to frame the family of standard explanations of the twin paradox by looking at the sorts of asymmetries that are claimed as would-be explanations for the twins' different ages.

Langevin himself pointed out two possible asymmetries in the experiences of the twins that foreshadow the two major approaches generally taken to explain the twins' ages.[19] These are as follows. First of all, Langevin noted the contrast between what each twin might detect by keeping track of the other using radio signals. In this way, each twin may "watch" the other age in such a way that their different ages are unsurprising. Second, Langevin pointed out that the traveling twin undergoes an acceleration as he turns back toward Earth, but the earthbound twin stays at rest in a single inertial frame. Thus one has an asymmetry. As already proposed, these two asymmetries form the basis for two families of subsequent explanations for the twins' different ages, those that seek to keep track of the twins' relative ages throughout the journey and those that look to the acceleration of the traveling twin as the key to understanding the problem.

Taking a closer look at the first of these families, one may note that in order to keep track of the twins' relative ages while they are apart, one must adopt some form of convention for distant synchrony. Thus, these accounts turn fundamentally on different simultaneity criteria. This may be illustrated by several different but related approaches. Among these explanations are the radio signal approach first suggested by Langevin and Lord Halsbury's "three brothers" approach. These both explicitly or implicitly remove from consideration the role of the acceleration. Each then tells a story about how during the course of the journey the proper times measured by earthbound and traveling clocks change with respect to one another.

David Bohm gives a detailed version of the radio signal approach. The different experiences he describes of traveler and earthbound observer while maintaining radio contact follows Langevin's qualitative discussion closely. See Figure 5.3. From the relativistic Doppler shift equations, Bohm notes that from the point of view of the earthbound observer, he or

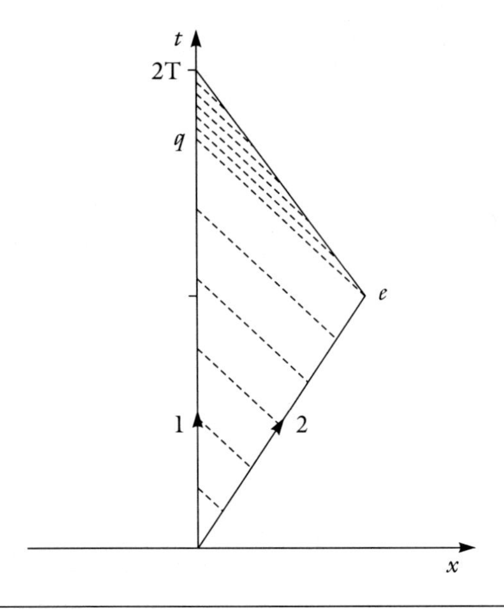

Figure 5.3. The dashed lines depict radio signals sent at constant temporal intervals from the frame of the traveling twin, path 2, back to Earth. The point q indicates the moment that the signal sent from turning point e is received by the earthbound twin on path 1.

she will receive "first of all a set of slower pulses and later [after time q], another set of faster ones,"[20] where $q = \mathrm{T}(1 + v/c)$ is the time that the first signal is received after the traveling twin turns around.

Conversely Bohm concludes that "If the rocket observer were watching the fixed observer he would then see the life of the latter slowed down at first and later speeded up."[21] The change between slow and fast would in this case occur at the time $p = \mathrm{T}(1 - v/c)$ when a signal from Earth reaches the traveler at the turnaround point e. See Figure 5.4. Bohm concludes that for the traveling twin "the effect of the speeding up more than balanced that of the slowing down. He would not therefore be surprised to find on meeting with his twin that the latter had experienced more of life than he had."[22] Bohm's account of the relative lapse of proper time for each observer does not give the acceleration any special treatment and describes a situation in which each observer sees the other going more slowly at first and going faster after a certain moment in time, p or q.

The radio communication solution exemplified by Bohm is similar to the "three brothers" approach suggested by Lord Halsbury. This is a situation in which instead of turning the corner at e, the traveling twin's clock

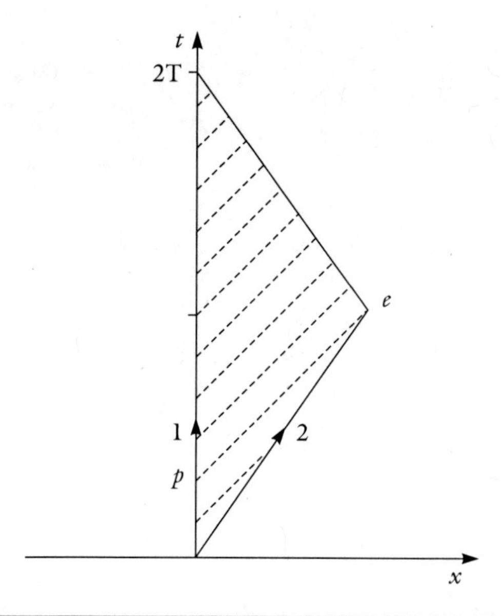

Figure 5.4. The dashed lines depict radio signals sent at constant temporal intervals from the frame of the earthbound twin, path 1, to the traveling twin. The point *p* indicates the moment of transmission of the signal from Earth to the turning point, *e*, on path 2.

is synchronized with a third clock carried by a third sibling already moving at the opposite velocity toward Earth; the time measure by both clocks will together give us the proper time along the whole of path 2.[23] This is intended to remove any question of the effect of acceleration on the motion.[24] The difference in measurements of proper times on the two paths, according to those who have adopted this approach, is (as in Bohm's discussion) based on the relativity of simultaneity. Each inertial frame—stationary, departing and returning—has lines of simultaneity, horizontal and parallel to line segments *re* and *se* respectively, each defined by the Einstein convention, as shown in Figure 5.5. Therefore on outgoing and returning legs, both traveling and stationary clocks seem to be going faster than each other, but the change of inertial frames at *e* constitutes a change of lines of simultaneity that results in a jump ahead between the times *r* and *s* as measured on the moving clocks with respect to the stationary clocks. The "missing time" between *r* and *s* becomes then the basis for explaining the differential aging.

What may be said about those accounts which assert that Langevin's second asymmetry, the direction-reversing acceleration, is essential to a

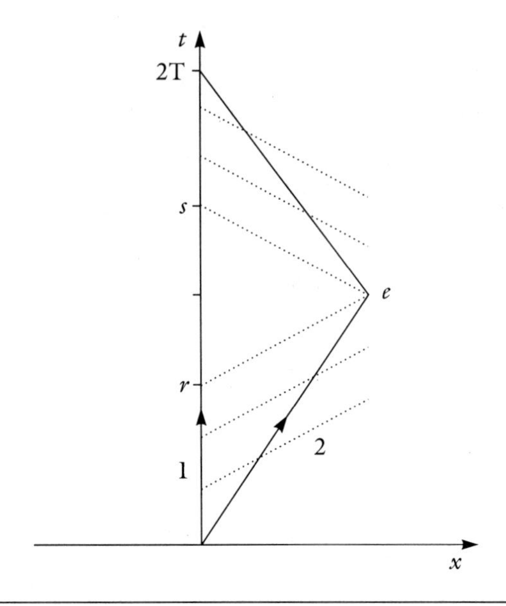

Figure 5.5. The interval *rs,* along the time axis, has sometimes been thought of as the "missing time" on the traveling twin's clock, brought about by a change in standards of synchrony (depicted as dotted lines).

complete explanation of the paradox? This is not the case. It is still possible to come up with a similar scenario in which both twins travel and accelerate upon turning back home, but in which differential aging still occurs; this may be done by having both twins travel but at different speeds. Using equations (5.1) and (5.2) to calculate proper time will lead to the conclusion that twins age differently, while clearly both twins undergo an acceleration. It should be obvious from this that the asymmetry of acceleration does not provide an explanation for the different ages of the twins.

Even so, many still feel that the introduction of general relativity and a gravitational field at the point of acceleration is the best way to explain this second asymmetry. Bohm, in fact, expresses this view and notes that "two clocks running at places of different gravitational potential will have different rates."[25] To cite another example, S. P. Boughn uses just this observation to interpret his explanation of differential aging in a version of the twin paradox.[26] In his paper, Boughn argues that "identically accelerated twins" also age differently and that this effect is due to the way that proper time is calculated in a uniform gravitational field. Since one is dealing with flat spacetime in special relativity, the reference to general relativity and gravitational fields in this context is misleading.[27]

A Generalized Scheme

The result of both simultaneity-based and acceleration-based explanations of the twin paradox has been a situation in which discussion centers on trying to say where and when the traveler loses time against Earth. However, one may develop an approach to the differential aging problem that promises to lay all of this discussion aside. In particular one may apply the conventionality of simultaneity, as advocated by Reichenbach and Grünbaum, to the twins problem.[28] Adopting this approach, one may incorporate various suggested accounts that seek to explain the different ages of the twins as specific cases of a generalized scheme.

Recalling the conventionality of simultaneity in using light signals to synchronize two spatially separated clocks at points *a* and *b* (shown in Figure 5.6), one need not divide the difference of transmission, t_1, and recep-

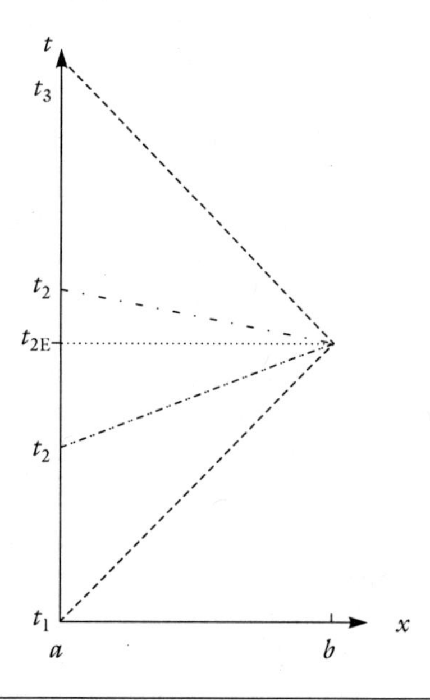

Figure 5.6. Establishing synchrony between spatially separated points *a* and *b*, in one spatial dimension. The dashed lines depict radio or light signals. The dotted line depicts the standard of synchrony associated with the Einstein convention. The mixed dashed and dotted lines depict other possible standards of synchrony.

tion, t_3, of a signal by two as originally described by Einstein. Doing so would give us Einstein's convention of simultaneity, represented by t_{2E}, which is equivalent to assuming the constancy of the one-way speed of light. Instead, one could choose any time, t_2, measured at position *a* between t_1 and t_3 to be simultaneous with the time of reception of the signal at position *b*. Another way of saying this is that the interval from t_1 to t_3 is topologically simultaneous with the time of reception recorded at *b*. As already noted, John Winnie has investigated some consequences of this approach to establishing simultaneity, using Reichenbach's notation:[29]

$$t_2 = t_1 + \varepsilon(t_3 - t_1) \tag{5.10}$$

such that $0 \leq \varepsilon \leq 1$. When $\varepsilon = \frac{1}{2}$, this is equivalent to the Einstein convention. Winnie points out that any simultaneity criterion, such that $0 \leq \varepsilon \leq 1$, may in fact be applied without affecting the differential aging in what is equivalent to a Halsbury-type phrasing of the twin paradox. This is what one would expect if the choice of ε is truly one of convention.

However, Winnie also concludes that the standard time dilation in special relativity described by the phrase "moving clocks run slow" is in some ways an artifact of the Einstein convention. Winnie calculates specific criteria, i.e., values of ε, according to which clocks can be seen to run synchronously in rest and moving frames. To do this and remove any one-way time dilation, he shows that one must choose different values of ε for when the clocks being synchronized are receding, ε_r, or approaching, ε_a, with respect to the rest frame, and that these values are additive inverses of each other, such that:

$$\varepsilon_r + \varepsilon_a = 1 \tag{5.11}$$

What Winnie's work suggests for the twin paradox is that whereas round-trip differential aging is not dependent on convention, the one-way description of relative clock rates is. Thus any choice of simultaneity criterion, ε, will yield the overall difference in age for the twins, but that each different choice will represent an equally acceptable story about the relative rates of clocks along each portion of the journey. If this is the case, then any of the discussions of where or when during his or her journey the traveler gains on the earthbound twin become equally conventional.

In order to apply the conventionality of simultaneity to the problem of relative rates of clocks in the twin paradox, consider a situation in which the traveling twin continuously sends and receives signals from Earth and uses these to set upper, *u*, and lower, *l*, bounds on possible values of his or her clock. See Figure 5.7. One may plot the proper time along each path against one another, which produces a parallelogram (as shown in Figure 5.8), the upper and lower boundaries of which are the bounds on possible times measured by the traveler for a given instant on the earthbound clock. That is to say that for each instant of proper time measured along path 1, τ_1, there exists a range of proper time values along path 2, τ_2, which would all be equally good choices to be considered simultaneous with that particular value of τ_1. It is important to note that the diagonal of the parallelogram lies below the line of slope equals 1 such that the differential aging by the end of the journey is undisputed. This parallelogram, OPQR, also allows one to see that it would be possible for either clock to run more quickly than the other over any particular interval within its bounds. The essence of the conventionality of simultaneity approach to the twin paradox can be made apparent by remarking that any nondecreasing curve inside the parallelogram would be equally acceptable as a way of describing the relative rates of the two clocks.

With this approach to the twin paradox it is easy to see that any discussion about where during the journey the differential aging takes place is

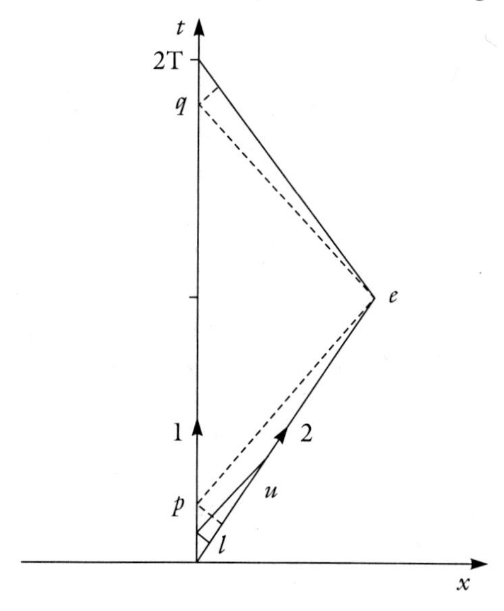

Figure 5.7. The dashed lines depict radio or light signals. The bounds on possible simultaneous points are given by the proper times measured on path 2 at each point of transmission, *l*, for the lower bound, and reception, *u*, for the upper bound.

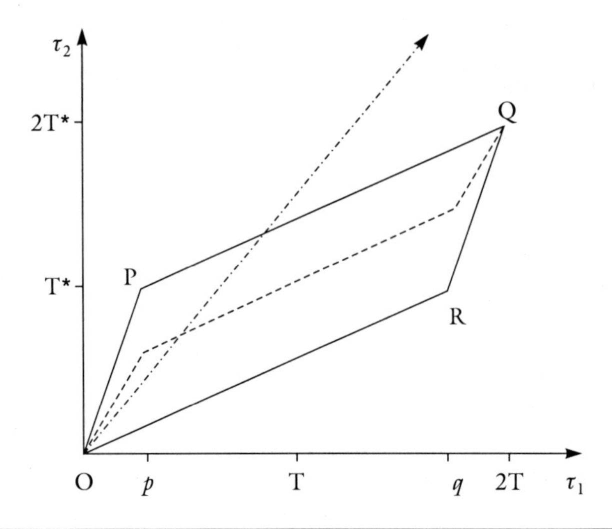

Figure 5.8. A parallelogram of possibly simultaneous points. The dashed line depicts one possible simultaneity convention. $T^* = T\gamma^{-1}$. The mixed dashed and dotted line indicates the line of slope = 1, on which $\tau_1 = \tau_2$.

unnecessary. In fact, many of the standard explanations can be plotted onto the parallelogram. Two of those already discussed involve the use of the Einstein convention from the point of view of the traveler. In the first, one can simply halve the difference between the traveler's sending and receiving times over the whole journey to establish the progress of the twins' clocks relative to one another. Graphically this would mean taking the average between the upper and lower boundaries of the parallelogram, as depicted by the dashed line segments in Figure 5.8. Adopting this method, the traveler's clock seems to run more quickly than Earth's up to $\tau_1 = T(1 - v/c)$, the time in Earth's frame that the first signal reaches the turnaround point. Then the traveler's clock seems to lose ground against Earth's until $\tau_1 = T(1 + v/c)$, the time that Earth receives its first signal after the turnaround. Thereafter the traveler again ages more quickly but the overall effect is such that his or her total age is less than that recorded on Earth.

The Halsbury "three brothers" approach to explaining the differential aging can also be represented as a curve inside the parallelogram, shown by the dashed line segments in Figure 5.9. To get this curve, the Einstein convention for simultaneity for a frame receding with velocity v is used un-

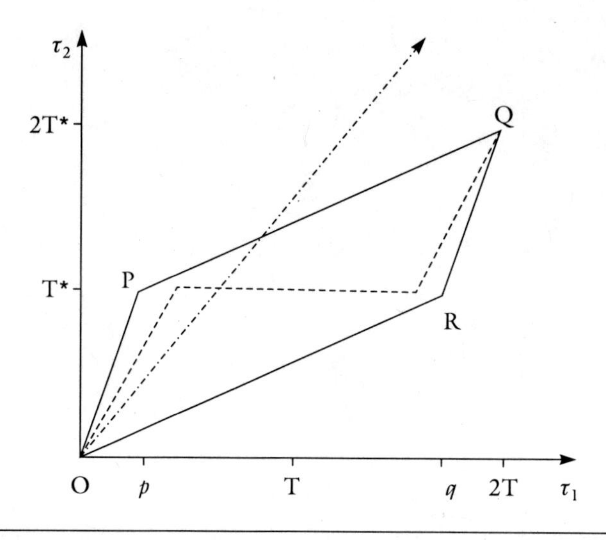

Figure 5.9. The dashed line segments depict a possible choice of simultaneity conventions corresponding to the Halsbury "three brothers" approach to the twin paradox.

til the traveler reaches his or her halfway point, half of τ_2, implying that the traveler is aging more quickly. On the second half of τ_2, the same convention for an approaching frame is used, and the traveler ages more quickly again. The overall youth of the traveler is due then to the "missing time" from the traveler's journey that is represented by the horizontal section in which Earth ages instantaneously from his or her point of view. Conversely, from Earth, the traveling twin's clock seems to stand still during this period.

Infinitely many other stories may also be told that fit into the bounds of convention set by the parallelogram. As previously shown, the simultaneity criterion, ε, can be chosen so as to eliminate one-way time dilation if $\varepsilon = \varepsilon_r$, for receding clocks, or ε_a, for approaching ones. The result of choosing these criteria is represented by the dashed line segments in Figure 5.10. During the first half of the traveling twin's journey, his or her clock runs in synchrony with the clock on Earth, the dashed line segment runs along the line of slope 1. These lines are also parallel during the second half of the twin's journey, where the clocks run at the same rates again. The overall differential aging is caused by the horizontal dashed segment over which, from Earth, the traveler's clock stands still. Interest-

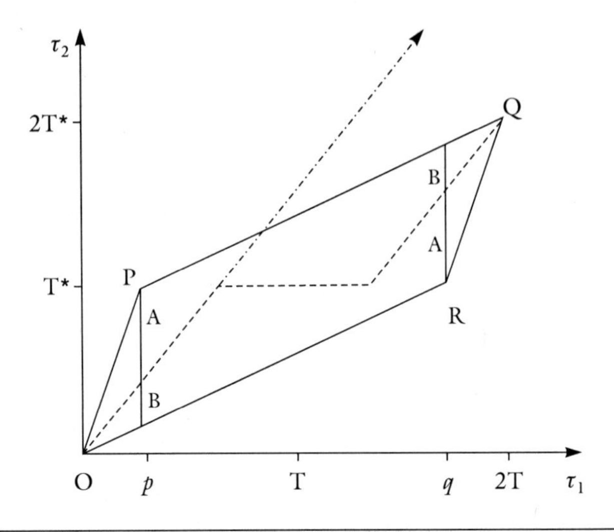

Figure 5.10. The dashed line segments depict another possible choice of simultaneity conventions.

ingly, the additive inverse relationship of expression (5.11) can be seen readily on the parallelogram. Modifying Reichenbach's notation in (5.10):

$$\varepsilon = (t_2 - t_1)/(t_3 - t_1). \tag{5.12}$$

From the traveler's point of view, t_3 is equal to the upper boundary of the parallelogram, t_1 the lower boundary, and t_2 the chosen simultaneous moment represented by the dashed line segments. This implies that on the first half of the traveler's journey,

$$\varepsilon_r = B/(A + B) \tag{5.13}$$

where A and B are the magnitudes labeled on Figure 5.10. On the second half of the journey,

$$\varepsilon_a = A/(A + B). \tag{5.14}$$

This gives the result that

$$\varepsilon_r + \varepsilon_a = (A + B)/(A + B) = 1, \tag{5.15}$$

as expected.

The boundaries of the parallelogram can also be seen to represent the approach to explaining the twin paradox, exemplified above in Bohm's discussion, which uses Doppler shifted radio signals. Looking more closely at the boundaries of the parallelogram, one can see that the lower boundary has the slope $[(1 - v/c)/(1 + v/c)]^{1/2}$ and the upper boundary has the slope $[(1 + v/c)/(1 - v/c)]^{1/2}$ where v is taken to be the outgoing velocity of the traveler and $-v$ the returning velocity. This is not surprising as these slopes are the relative rates of the measurement of proper times in frames moving with respect to one another. This relationship can be seen in the relativistic Doppler shift equation according to which

$$\tau' = [(1 + v/c)/(1 - v/c)]^{1/2}\,\tau \qquad (5.16)$$

where τ' is the period of radiation received in a frame moving with velocity v and τ is the period of the radiation in the rest frame, in the situation where the radiation is propagating in the same direction as v. Noting that $\tau_2 = \Sigma\,\tau'$ and $\tau_1 = \Sigma\,\tau$ over their respective paths, and that the Doppler shift equation describes the periods of radiation on the upper bound and the multiplicative inverse describes the lower bound, the slopes of the sides of the parallelogram can be easily confirmed.

Looking to the story Bohm tells of the twins' relative progress, one can see that he is actually describing the two boundaries of the parallelogram that come directly from the Doppler-shifted signals he sets out to discuss. Bohm first discusses the appearance of signals coming from the traveler as seen on Earth. Looking at the parallelogram, Bohm's explanation corresponds to the lower boundary, ORQ, in Figure 5.10.[30] Taking this approach, the Earth observer sees the traveling twin aging more slowly up to the time $p = T(1 + v/c)$, represented by segment OR of the parallelogram. Subsequently he or she sees the traveler aging more quickly than earthbound clocks, segment RQ.

From the other point of view, Bohm expects that the moving twin will see the Earth's clock running slower than the moving clock up until the time $q = T(1 - v/c)$, and subsequently he or she will see the Earth's clock running more quickly than the moving clock.[31] This is exactly the story that is represented by the upper boundary, OPQ, of the parallelogram in Figure 5.10. Bohm explains the differential aging by pointing out that the speeding up of Earth's clock witnessed by the traveler after time q "more than balanced" the slower relative rate prior to q.[32] The use of the parallelogram makes it obvious that the Doppler-shifted signal approach to the paradox is concerned with the outer bounds of an infinite number of acceptable stories about the twins' relative rates of aging.

The Twins on Nonstandard Paths

In addition to the various attempts to explain the standard version of the twin paradox, some have sought to extend the discussion by considering nonstandard paths; the generalized scheme is also able to deal with these cases. One such approach to explaining the differential aging of the twins without reference to the point of acceleration has been to put the two paths onto cylindrical coordinates. Cylindrical 1 + 1 dimensional spacetime may be thought of as a two-dimensional universe in the shape of an infinitely long cylinder with time running "up" the cylinder and space running "around" it. In this way the stationary twin is considered to travel up the cylinder, the time axis parallel to the axis of rotation of the cylinder, and the traveling twin to depart and return by simply going around the cylinder at a constant velocity. The proper times on the cylinder have been calculated recently by more than one individual.[33] At first it might seem that one might be able to get a real paradox out of this situation without the obvious asymmetry in the two paths provided by the acceleration and change in direction on the traveler's path. This turns out not to be possible because of the structure of simultaneity relations in cylindrical spacetime.

Another nonstandard version of the twin paradox appears in discussions of the Hafele-Keating experiment. In this experiment, differential aging was observed on two atomic clocks traveling on jets at the same speed around Earth in opposite directions.[34] These two paths (without accounting for the rotation of Earth) can be compared schematically to the two paths going around in different directions on cylindrical coordinates. The addition of the rotation gives us a scenario that looks roughly like the twin paradox on cylindrical coordinates.

Using flat noncylindrical coordinates one can set up an idealized one-dimensional Hafele-Keating situation with two symmetrical paths for each twin corresponding to the case when Earth's rotation is not considered. See Figure 5.11. It turns out that using the Einstein convention of simultaneity implies a specific story about the relative rates of clocks even when there is no overall differential aging. Taking the same approach as above, one can consider that the twin on path 2 checks the progress of the other by sending and receiving signals. With this information, the twin on path 2 again sets upper and lower bounds for acceptable values of its own proper time, τ_2, for each given value of the other twin's proper time, τ_1. The result is a symmetrical hexagon with its diagonal along the line of slope 1, so that there is no overall differential aging. See Figure 5.12.

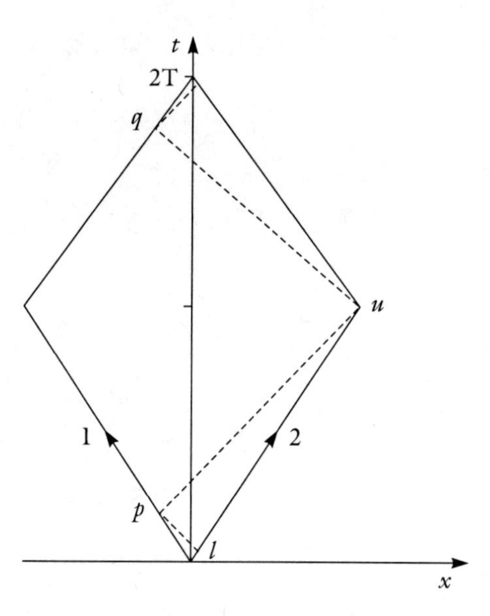

Figure 5.11. A 1 + 1 dimensional "Hafele-Keating" scenario. The labeled arrows designate the paths of the two twins. The symmetrical paths do not take into account Earth's rotation; there is no differential aging. The dashed lines depict radio or light signals. The point q indicates the moment that the signal sent from the turning point on path 2 is received by the twin on path 1. The point p indicates the moment of transmission of the signal from the twin on path 1 to the turning point on path 2.

The slopes of the sides of the hexagon can once again be explained using the Doppler shift equation (5.16), according to which, as previously stated, the lower boundary has the slope $[(1 - v_r/c)/(1 + v_r/c)]^{1/2}$ and the upper boundary has the slope $[(1 + v_r/c)/(1 - v_r/c)]^{1/2}$, where this time v_r is the relative velocity between the twins, each moving with velocity v. Using relativistic velocity addition:

$$v_r = 2v/(1 + v^2/c^2) \tag{5.17}$$

from $\tau_1 = 0$ to p, where $p = [T - T(2v/c)/(1 + v/c)]\gamma^{-1}$, which corresponds to segments OP and OT in Figure 5.12. Plugging into the Doppler shift equation gives slopes for these segments of $(1 + v/c)/(1 - v/c)$ and $(1 - v/c)/(1 + v/c)$, respectively. The relative velocity, v_r, is zero from $\tau_1 = p$ to q, where $q = [T + T(2v/c)/(1 + v/c)]\gamma^{-1}$, which implies that segments PQ and TS are both of slope 1. The other boundaries are similarly calculated.

The dashed line segments inside the hexagon represent the story given

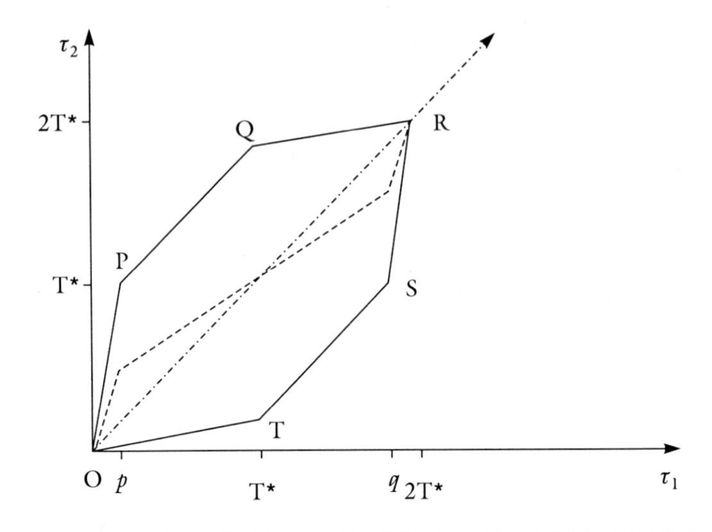

Figure 5.12. A symmetrical hexagon of possibly simultaneous points. The dashed line depicts one possible simultaneity convention. $T^* = T\gamma^{-1}$. The mixed dashed and dotted line indicates the line of slope = 1, on which $\tau_1 = \tau_2$.

if the Einstein convention is used in the sense that the average of the upper and lower bounds is taken everywhere. As is evident from Figure 5.12, use of this convention implies that the clocks on different paths are seen to move faster or slower than each other at different moments during the journey even when there is no overall differential aging. In fact no single simultaneity criterion, ε, will pick out the diagonal that seems to make the most sense as a description of the clocks in this situation. The arbitrary nature of the implications of any single criterion supports the conventionality approach of setting the boundaries and not specifying paths within them.

One can also create a one-dimensional Hafele-Keating experiment in which there is differential aging by adding a velocity in one direction (see Figure 5.13). In this situation the magnitude of the velocities on path 1 is less than that of the velocities on path 2, and the overall aging of the twin on path 1 will be greater. A hexagon can also be drawn to incorporate the possible values of one proper time versus the other, as shown in Figure 5.14. The slopes of the sides, as in the previous hexagon, can be given using the relativistic Doppler shift equation (5.16) and the relative velocities of the two twins using relativistic addition of velocities. As one would expect, substitution of the same velocity for paths 1 and 2 produces the sym-

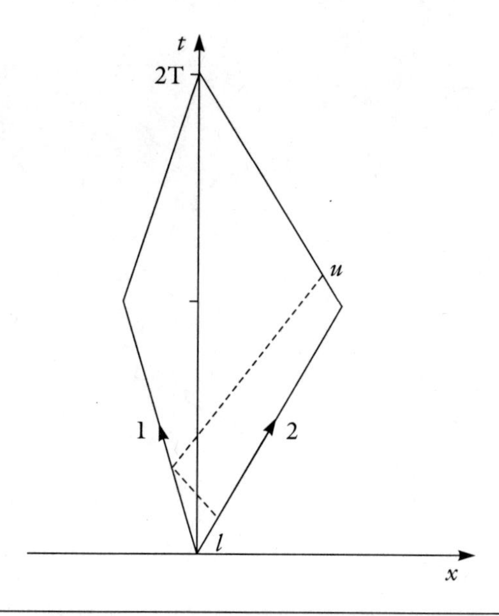

Figure 5.13. A 1 + 1 dimensional "Hafele-Keating" scenario. The labeled arrows designate the paths of the two twins. Nonsymmetrical paths correspond to the case in which Earth's rotation is considered; the twins age differently.

metric hexagon in Figure 5.12, and substitution of zero velocity for one of the paths produces the parallelogram from the standard twin paradox; see Figure 5.8.

Some general features of this approach to depicting the relative progress of clocks between two paths in Minkowski spacetime can be observed. First of all, it is possible to construct a region of possibly simultaneous points for any two paths. The Doppler shift equation relating periods of signals can be used to sum over all periods to get the upper and lower bounds on this region as long as relative velocity along each path is constant over each individual period. Much simpler methods can be used to calculate these bounds if the paths are straight and accelerations are instantaneous. In this situation, one can see from the examples presented so far that the number of vertices, V, on the boundary of the simultaneity region is given by

$$V = 2(n + 1) \tag{5.18}$$

where n is the number of instantaneous points of acceleration on the twins' paths, excluding the accelerations at separation and return.

The conventionality of simultaneity approach to the twin paradox also

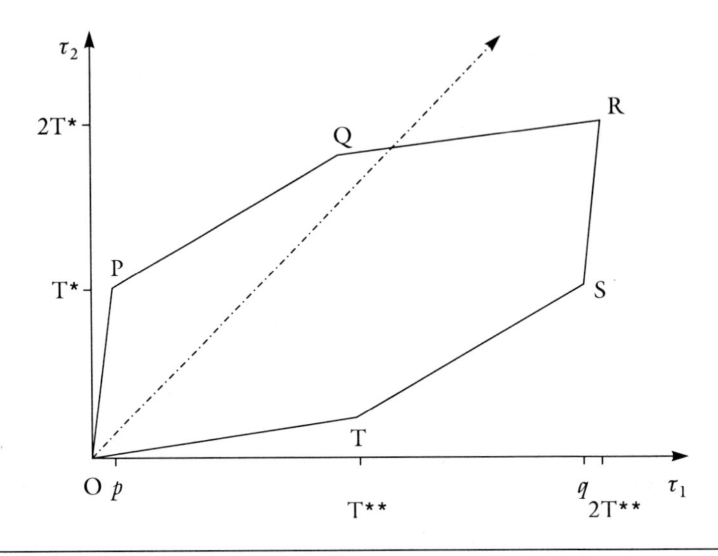

Figure 5.14. A nonsymmetrical hexagon of possibly simultaneous points. $T^{**} = T\gamma^{-1}{}_1$ and $T^{*} = T\gamma^{-1}{}_2$, where $\gamma^{-1}{}_1$ and $\gamma^{-1}{}_2$ are different on paths 1 and 2 respectively. The mixed dashed and dotted line indicates the line of slope = 1, on which $\tau_1 = \tau_2$.

clarifies some implications of a method for estimating distance suggested by Clive Kilmister. Kilmister has suggested that the traveler could keep track of his or her distance from the origin of the rest frame using the same signals used above to discuss simultaneity.[35] By this method, described as a "radar" method by Hermann Bondi, the traveling twin could estimate distance from Earth by estimating the time it takes for a signal to make the trip using the Einstein convention and multiplying by the speed of light.[36] This is equivalent to saying that the distance, *d*, for a specific value of τ_1, the proper time measured on path 1, is given by

$$d = c(t_u - t_l)/2 \qquad (5.19)$$

where t_u and t_l are the upper and lower bounds on the value of proper time on path 2, for that value of τ_1. From the parallelogram and hexagons already constructed one can get an idea of how the quantity $(t_u - t_l)/2$ varies at different proper times τ_1. Multiplying by c results in the distance estimate of expression (5.19). For the standard twin paradox situation, the radar method implies that the distance between twins is con-

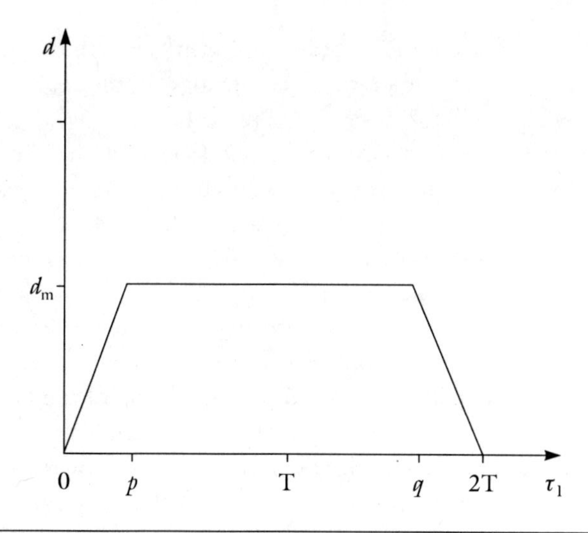

Figure 5.15. Distance estimate by the radar method versus proper time on path 1; the estimate levels off at d_m.

stant at a maximum distance, d_m, near the change of direction. See Figure 5.15. In the symmetrical one-dimensional Hafele-Keating situation, the radar method also gives an artificially low estimate of distance which implies that the relative velocity between the twins is lessened near the turning points. The distortion of these distance estimates is a result of the use of the radar method and the specific simultaneity criterion it assumes, and these strange results are artifacts of its adoption. This example demonstrates, this time from the point of view of relative distance instead of relative aging, the arbitrary results of choosing a single simultaneity criterion.

The Role of Acceleration Put to Rest

Finally, this generalized scheme can also be used to show once more that discussions that attribute the age difference between the twins to the direction-reversing acceleration are beside the point. One recent example of such a discussion is that between S. P. Boughn and co-authors Edward A. Desloge and R. J. Philpott.[37] This interchange seems to have been inspired by the idea that two twins that undergo the same acceleration may age differently as a result. This was cited by Boughn as providing an "important insight into the behavior of clocks in a uniform gravitational field," with a mind to application to the point of acceleration in the standard formulation of the twin paradox. Desloge and Philpott have responded by describing in more detail the paths on Minkowski spacetime that Boughn's scenario requires, if the journeys of the twins are to start and finish in spatial coincidence. A version similar to that which they describe is pictured in Figure 5.16. In this case the twins are separated symmetrically, given the same acceleration into a new frame at a point *e* in time, and brought back together symmetrically with respect to their new frame. At the end, the twin on path 2 has a greater total elapsed proper time.

Applying the generalized approach to this version of the twin paradox, one could in principle draw a region of possible simultaneous points that would have fourteen vertices—see equation (5.18). Any path within this region would be an acceptable account of the differential aging. One need not construct the entire region to see that assigning the difference in age to the point of acceleration is only one of these accounts. In fact, one can see this from a rough diagram of the shape of this region around the point

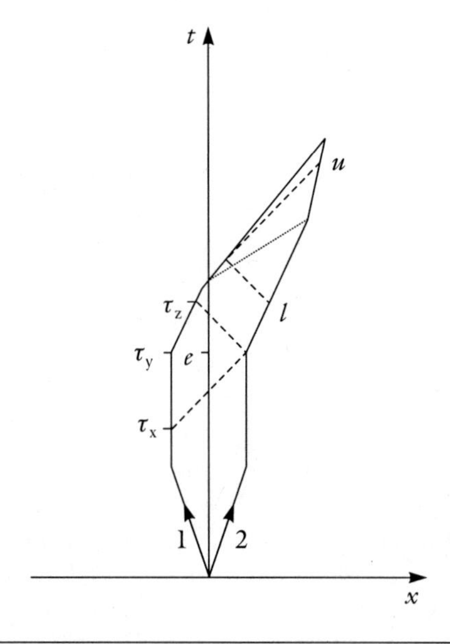

Figure 5.16. An "identically accelerated twins" version of the twin paradox. The labeled arrows designate the paths of the two twins. The dashed lines depict radio or light signals. τ_x, τ_y, τ_z, are points in proper time measured on path 1 near the moment of acceleration.

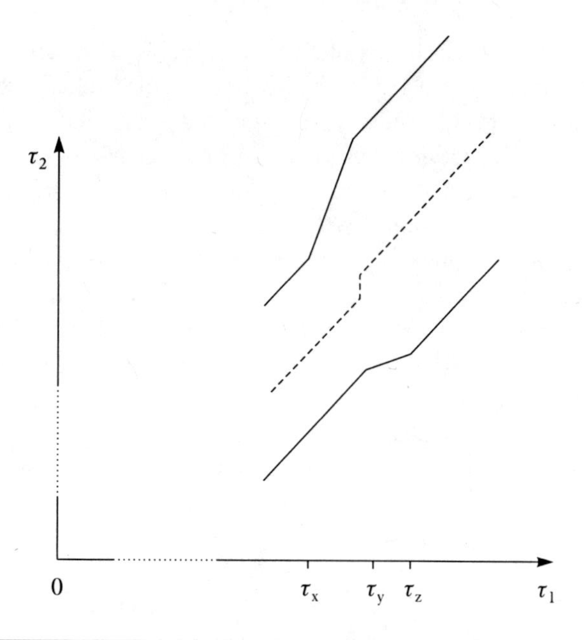

Figure 5.17. An approximate detail of the region of possibly simultaneous points, near the moment of acceleration. The dashed line segments indicate a possible choice of simultaneity conventions.

of acceleration (see Figure 5.17). The dashed line segments designate a story that allows the differential aging to take place at the point of acceleration. However, it is obvious that many other nondecreasing curves could fit within the appropriate bounds.

It remains true, of course, that without acceleration (of at least one of the twins), then in Minkowski spacetime, at any rate, it is impossible to have twice-intersecting trajectories so as to formulate the twin paradox with the twins starting and finishing in spatial coincidence. So in this sense acceleration is an essential ingredient in understanding the twin paradox. It may be noted, however, that even this role for acceleration can be eliminated in formulations of the twin paradox in curved spacetime.[38]

Implications of the Generalized Scheme

On the basis of this generalized scheme for explaining the twin paradox, one may conclude that any explanation of the twins' different ages that depends on specific implicit or explicit simultaneity conventions is just that, conventional, as it discusses only one of an infinite number of conventional ways to approach the problem. In perspectival invariantist terms any one of these explanations fails to be objective. This is due to the fact that specific simultaneity conventions are associated with specific sets of inertial coordinates, and as previously shown, pure spatial and temporal intervals that make up these sets of coordinates are not invariant under the Poincaré group, the relevant invariance criterion for special relativity. Additionally, pure spatial distances also fail to be invariant, as in the example above of Bondi's radar method.

Thus the various retellings of the twin paradox seek to explain the different ages of the twins by appeal to representational structures that are not legitimately part of the formal chain of media of an objective representation. From a perspectival invariantist perspective, the only representation that does come out as objective is one that does not include specific sets of coordinates (based on specific simultaneity conventions). An example of an objective representation of the twin paradox is the minimal standard version presented at the start of this chapter. This is not to say that one may not make use of specific coordinate frames in an objective representation. In fact, such a frame is a necessary part of the proper time-calculation procedure. Thus it helps one establish an invariant representational relation, but it is not invariant itself and therefore not part of the formal chain of media of an objective representation.

Time and Perspectival Invariance

The case of the twin paradox illustrates two central features of a perspectival invariantist approach. First of all, it demonstrates how group theoretical invariance, via the notion of perspectival invariantism, may be used to pick out representations that are the basis for better scientific explanations. Second, one can see that invariance also indicates the scope available for conventional choice; there is no objective fact-of-the-matter to distinguish any story of the twins that does violate the invariance requirement.

In the case of the twins, perspectival invariantism picks out a family of representations that share a representational relation, the clock hypothesis, and an invariance criterion, the Poincaré group. The representational relation between the elapsed time measured on a clock, t_c, and the proper time, τ, measured along the path of that clock, is between two structures that remain invariant under the Poincaré group. Because other structures, including coordinate frames, spatial, or temporal intervals, are not invariant under this group, then they may not be used as part of the formal chain of media linking elapsed time with proper time. Since the usual explanations of the twins' ages make use of one noninvariant structure or another, then one may reject them as nonobjective, as above. Thus perspectival invariantism picks out the representations of the twin paradox that are the basis for better explanations of the twins' different ages.

It might seem, however, that in restricting ourselves to objective (Obj_P) representations, one might have removed the conventional ambiguity that is inherent in the notion of simultaneity in special relativity. This is not the case. The absolute conventional ambiguity over distant clock synchrony is still very much a feature of the generalized account of the twin paradox presented above. At the same time, through perspectival invariantism, one is able to provide an explanation for the twins' ages that is not vulnerable to this ambiguity.

This case study therefore stands as an example of how representation in physics may be at once conventional and objective when the invariance criterion and representational relation are specified (in this case the Poincaré group and the clock hypothesis, respectively). Thus, if invariance is understood as a necessary and sufficient condition for objectivity, then it is equally a key to the scope of conventional choices available.

6

Localization in Quantum Theory

The final case considered here is perhaps unexpected as an exemplar of perspectival invariantism. As previously mentioned, the proponent of this trope of invariantism seeks to salvage a necessary condition out of the claim that objectivity means invariance by restricting the relevant class of symmetries to those that can be associated with actual or idealized points of view, on the one hand, and those that are generalizable, on the other. This case concerns the representation of quantum mechanical systems, and an observer's perspective on a state vector might at first seem difficult to conceptualize. This is evident in reference to the extensive body of literature that addresses the relationship between quantum mechanics and the relativistic understanding of the spacetime in which these systems live. Of particular interest here, of course, is the set of spacetime symmetries associated with special relativity.

The particular point at issue is the representation of a single particle located at a specific position in space within the formalism of so-called "relativistic quantum mechanics."[1] In perspectival invariantist terms, the relevant representational relation here should exist between a particle and a quantum state vector, and the invariance criterion (at least initially) is once more the Poincaré group of special relativity. As will become apparent, problems arise for quantum mechanical localization when combined with the relativistic consideration of moving frames of reference. These problems lead to the failure of this representational relation to remain invariant under the Poincaré group. Thus one is faced with a choice between altering either the representational relation or the invariance criterion. It will be shown that either of these options can lead to a different objective *(Obj_P)* representation, either by adopting a quantum field-theoretical approach or the so-called "hyperplane-dependent" formalism proposed by Gordon Fleming.

Position and Localization

Almost all of existing physical theory rests, at least in part, on a concept of spatial position. At the very least, this reveals that theorists have had a strong intuition that position should be a central aspect of our representations of reality. Whatever its origin, however, it is significant that a theoretical tradition that rests on well-defined spatial coordinates can serve to question the objectivity of this very aspect of the description of physical systems. Relativistic quantum mechanics provides an example of such a theory, one in which the property of being located at a particular position in space is not invariant under the relevant invariance criterion of relativity theory, the Poincaré group. Before discussing this example in detail, one may note some of the conceptual problems related to the notion of position.

Problems with an intuitive concept of position, which is both arbitrarily precise and invariant across the relevant transformations, are not peculiar to relativistic quantum mechanics. The difficulty caused by length contraction, the noninvariance of pure spatial intervals in special relativity, has already been discussed in the previous chapter. As it turns out, the invariant quantity along a worldline is a mixture of spatial and temporal intervals that accompanies the conceptual shift toward linking the two into spacetime. As invariants, intervals of proper time, not distance, are the obvious candidates for use in objective representation within special relativity.

In nonrelativistic quantum mechanics, the concept of spatial position is captured by both spatial parameters and the dynamical variable of position represented by a self-adjoint operator; this distinction can also be made for the time parameter and an associated time-indicating dynamical variable.[2] Unlike the spacetime parameters, these dynamical variables are characterized by uncertainty relations that limit the sharpness with which a system's position can be determined in certain circumstances. In this case there may be in some sense no fact-of-the-matter as to where the system is precisely located.

Nonlocality in Nonrelativistic Quantum Mechanics

In addition to the problem of position uncertainty relations, the problem of nonlocality in quantum theory is one of the definitive conceptual problems in the interpretation of modern physics. This has traditionally been noted even within the discussion of nonrelativistic quantum theory. The well-known arguments include most famously those presented by Einstein, Podolsky, and Rosen (the EPR argument) and subsequently by John Bell in his eponymous inequality.[3] These have been analyzed and debated extensively both within physics and philosophy. One of the authors has summarized the conclusion of these debates in that "some sort of action-at-a-distance . . . seems built into any reasonable attempt to understand the quantum view of reality."[4] By action-at-a-distance, or nonlocality, he intends the violation of a locality principle termed "environmental locality," according to which "the value possessed by a local observable cannot be changed by altering the arrangement of a remote piece of apparatus which forms part of the measurement context for the combined system."[5] A remote piece of apparatus is one that is, in terms of special relativity, space-like separated from the local observable under consideration. It is evident from this that the whole discussion of quantum locality is essentially about how to embed quantum models into the representation of spacetime provided by relativity theory. Nonlocality, where it is evident, belies a fundamental tension between quantum mechanics and relativity.

Perhaps the most familiar context for the discussion of nonlocality is that set forth in the EPR argument. One may consider the well-known version of this argument in which one considers two spin ½ particles in a singlet state, measured at space-like separated wings of an experimental apparatus.[6] The purported nonlocal effect takes place when a measurement is taken on one wing of the apparatus, implying that whatever the outcome, say spin = +½, the particle on the other wing of the apparatus must be in an eigenstate associated with the opposite outcome, spin = −½. Moreover, this second piece of information comes into being instantaneously and thus constitutes a nonlocal effect.

The philosophical significance of the theoretical apparatus used to construct the EPR thought experiment turns on a few important concepts. These include notions of locality, elements of reality, and various possible realist or anti-realist interpretations of quantum objects; these are the representational structures associated with quantum systems. The various dilemmas implied by different combinations of positions on these points have been studied thoroughly; when combined with consideration of Bell's inequality, this leads to the conclusion already mentioned that nonlocality is an unavoidable feature of quantum theory, regardless of one's approach to the reality of quantum structures.[7]

One recent discussion of quantum nonlocality by P. Mittelstaedt has used the suggestive terminology of "objectification-at-a-distance" to refer to the action-at-a-distance inherent in EPR-like scenarios.[8] In this sense the eigenstate of the particle at the remote wing can be thought of as a quantum theoretical structure brought into existence by the measurement on the proximate wing. Mittelstaedt is attempting to keep open the possibility that this describes the objectification of an actual physical structure, a basis for a superluminal signal.

Without rehearsing these familiar arguments in their entirety, some features of the standard discussion are particularly relevant to the theme of representation in physics. First of all, it should be noted that the motivation of the authors Einstein, Podolsky, and Rosen was to provide what has been called an argument for the so-called "incompleteness" of (nonrelativistic) quantum mechanics. This idea of incompleteness has everything to do with the practice of linking models to structures in physical reality. As presented by one of the authors, the EPR argument begins with a "Necessary Condition for Completeness," according to which, "Every element of the physical reality must have a counterpart in the physical theory."[9] The idea here is simply to require that the theoretical language have a sufficiently rich structure to account for the features of reality it is to be used to represent. Equivalently, in the context of scientific representation, this amounts to a constraint on the chain of media and the representational relation between model and original. In other words, the idealized model O must not have any features that are not in principle representable by the mathematical model M.

In addition to the condition for completeness of a theory, the EPR argument depends on the requirement that a theory should obey a locality principle consistent with relativity theory. In order to obey this principle, causal influences must not propagate faster than light. After measurement then, the particle state at the remote wing of the apparatus must in fact have existed before the measurement took place. Since, according to basic quantum formalism, the quantum structure associated with this element of reality is only brought into existence at the instant of measurement, then there is an element of reality that is not captured by the representation; hence the purported incompleteness of quantum mechanics.

This incompleteness argument, although described in terms of elements of reality, is not primarily the result of experiment. It relates instead to representational structures within the EPR thought experiment, requiring that they must exist within the constraints of relativity theory. Within the formalism of nonrelativistic quantum mechanics alone the action-at-a-distance that takes place is perfectly acceptable. However, the nonrelativistic quantum state (eigenstate of spin) at the remote end of the apparatus is not one that satisfies the constraints of special relativity. One of these constraints is the prohibition on faster-than-light signaling.

The tension between relativity and quantum mechanics over the issue of nonlocality is usually resolved by resorting to the claim that even when superluminal—faster than light—phenomena are present, they cannot be used for signaling. This, it has been argued, allows for "peaceful coexistence" between quantum mechanics and relativity theory.[10] However, the claim that EPR-like correlations may not be used for superluminal signaling has been challenged. Mittelstaedt, for example, critiques various arguments in support of this thesis, suggesting that they are "merely plausible, but not really stringent and convincing."[11] Perhaps best known among the evidence for true superluminal propagation are the so-called "tunneling" phenomena. If indeed superluminal signaling occurs, then the peaceful coexistence between quantum mechanics and relativity would be undermined. Many have wondered, however, if all this concern over the disagreement between quantum mechanics and relativity might be done away with by looking at relativistic representations of quantum states.

Relativistic Nonlocality

If quantum mechanics in its nonrelativistic form unavoidably implies some sort of nonlocality, then what happens in explicitly relativistic treatments? There have been notable attempts to demonstrate that a relativistic formulation of the EPR argument does not lead to an action-at-a-distance that is at odds with special relativity.[12]

However, the example of nonlocality of primary interest in this chapter comes from a relativistic treatment, this time not of EPR correlations per se, but of localization in so-called "relativistic quantum mechanics." It is significant to note here that most researchers within the mainstream of current physics have concluded that there is no such thing as a relativistic quantum mechanics of localized particles.[13] This is doubly significant since it is precisely the nonlocality implied by localization in relativistic quantum mechanics that has led many to reject the theory ultimately in favor of quantum field theory.

In the case of relativistic quantum mechanics, the problem of nonlocality presents itself in the infinitely fast spreading of a localized wave packet. In a 1974 paper, Gerhard C. Hegerfeldt expressed the theorem that "if one insists on strict localizability [of a single particle state] . . . then causality will be violated."[14] His definition of causality is one that is essentially linked to locality, which is in turn based on the assumption that a particle may not propagate faster than the speed of light. Hegerfeldt's approach may also be expressed in the claim that "it is impossible to prepare a one-particle state which is strictly localized in a given finite space region V," on pain of violating relativity.[15] In the scenario he considers, a single particle state is localized at a given point in space; upon evolving the state forward in time, one finds that the probability of finding the particle indefinitely far away is not zero.

As will be demonstrated, Hegerfeldt nonlocality in relativistic quantum mechanics may be shown to be equivalent to the fact that the relevant localized quantum states are not invariant under Lorentz transformations; this phenomenon has recently been given the descriptive title of the "Jericho effect."[16] Since one may think of these states as representational structures and since Lorentz transformations are elements of the Poincaré group, then these states may not be part of an objective *(Obj_P)* representation. To the perspectival invariantist, this case suggests a link between the nonobjectivity of position, in the Jericho effect, and noninvariance of localization, as predicted by Hegerfeldt's theorem.

In the sections that follow, the link between Hegerfeldt's theorem and the Jericho effect is described through two methods of approximating the probability amplitudes predicted by each phenomenon respectively. This discussion will be framed by introducing localization in nonrelativistic quantum mechanics, relativistic wave-packets and the Newton-Wigner position eigenstate, and the difficulties associated with Newton-Wigner localization.

Localization as a Representational Relation

Before going on to this analysis, one may recall that one key element of scientific representation is the representational relation that links a model to an original (or a token to a type) via a formal chain of media. In the case of the twin paradox, this relation was that between elapsed time on a moving clock and the proper time calculated along its path through spacetime. In the case of representing localized quantum systems, the representational relation of interest is that between a quantum state vector and a single localized particle.

From the brief sketch of the notion of position and localization above one may note a couple of features of a quantum mechanical state vector used to represent a localized particle. First, the position value is sharply localized. In the terminology of quantum mechanics, this is equivalent to saying that the state is an eigenstate of the position observable. Second, the state vector must remain sharply localized under the relevant invariance criterion, the Poincaré group. With this in mind, one is in a position to consider the representation of localized quantum states in more technical detail.

Representing a Localized Particle

The issue of localization in relativistic quantum mechanics touches on many of the problems in the interpretation of modern physics. Practically, one would like to be able to find eigenstates the eigenvalues of which represent sharp spatial positions of a single particle. Even in the relatively straightforward case of nonrelativistic quantum mechanics, this is made somewhat problematic by the fact that infinitely sharp localization is best represented using Dirac delta functions. As is well known, these functions are formally known as generalized functions.[17] The fact that Dirac delta functions are treated as if they were functions in the ordinary sense throughout much of modern physics raises a number of important issues. However, as these important technical issues are not of primary interest here, Dirac delta functions will be used, as is consistent with standard practice.

Non-Relativistic Localization

In nonrelativistic quantum mechanics, localization can be expressed in terms of a (three-dimensional) delta function via the following inner product, where $\boldsymbol{\xi}$ and $\boldsymbol{\xi}'$ are two separate positions in space (bold variables indicate three-vectors).

$$\left\langle \Psi_{\xi'} | \Psi_{\xi} \right\rangle = \delta^3(\boldsymbol{\xi}' - \boldsymbol{\xi}) \tag{6.1}$$

In order to satisfy the conditions for quantum mechanical states, this inner product would need to be orthonormal. However, this is not the case above due to the fact that the delta function is not a normalizable function. Assuming, as above, that one may use Dirac delta functions as functions in the usual sense, this inner product may be used to define relative probability amplitudes.[18] Accordingly, one may treat them as if they were states in the formal sense. In terms of wave functions in **x**-space, this can be achieved in a straightforward way as follows:

$$\phi^{\xi}(\mathbf{x}) = \delta^{3}(\mathbf{x} - \boldsymbol{\xi})$$
$$\left\langle \phi^{\xi'} \middle| \phi^{\xi} \right\rangle = \int \phi^{\xi'*} \phi^{\xi} d^{3}x = \delta^{3}(\boldsymbol{\xi}' - \boldsymbol{\xi}).$$

The wave functions in momentum space can be constructed from this by taking its inverse Fourier transform.

$$\phi^{\xi}(\mathbf{k}) = (2\pi)^{-3/2} \int \delta^{3}(\mathbf{x} - \boldsymbol{\xi}) e^{-i\mathbf{k}\cdot\mathbf{x}} d^{3}x = (2\pi)^{-3/2} e^{-i\mathbf{k}\cdot\boldsymbol{\xi}'}$$

These position eigenstates, ϕ, are orthogonal to one another, but still not entirely adequate as eigenstates for the position of free particles as they are not orthonormal, since $\delta^3(0) = \infty$ when $\boldsymbol{\xi} = \boldsymbol{\xi}'$

A more general way to express the relationship between position and momentum states is to note that for a general quantum state $|\Psi\rangle$, its probability amplitudes with each are Fourier transforms of one another.[19]

$$\langle \mathbf{x} | \Psi \rangle = (2\pi)^{-3/2} \int \langle \mathbf{k} | \Psi \rangle \, e^{i\mathbf{k}\cdot\mathbf{x}} d^{3}k \tag{6.2}$$

This agrees with the eigenstates chosen above, but also allows for the introduction of a spreading function in order to ensure a finite norm in the position representation.

Proceeding in this manner, the wave function, $\left\langle \mathbf{x} \middle| \Psi^{\xi} \right\rangle \equiv \Psi^{\xi}(\mathbf{x})$, can be expressed as an integral over the original delta functions multiplied by a spreading function, F_{ξ} a Gaussian, such that:

$$\Psi^{\xi}(\mathbf{x}) = \int F_{\xi}(\boldsymbol{\xi}') \phi^{\xi'}(\mathbf{x}) d^{3}\xi' = F_{\xi}(\mathbf{x}) \tag{6.3}$$

Now taking the Fourier transform of this one can get an expression for the wave function where $\hat{F}_{\xi}(\mathbf{k})$, also a Gaussian, is the Fourier transform of the spreading function F_{ξ}.

$$\Psi^{\xi}(\mathbf{x}) = (2\pi)^{-3/2} \int \hat{F}_{\xi}(\mathbf{k}) e^{i\mathbf{k}\cdot\mathbf{x}} d^{3}k \tag{6.4}$$

With this expression for $\Psi^{\xi}(\mathbf{x})$ one can apply Parseval's theorem,

$$\int f^{*}(\mathbf{x}) f(\mathbf{x}) d^{3}x = \int \hat{f}^{*}(\mathbf{k}) \hat{f}(\mathbf{k}) d^{3}k \tag{6.5}$$

to get a finite norm as follows:

$$\left\| \Psi^{\xi}(\mathbf{x}) \right\|^{2} = \langle \Psi | \Psi \rangle = \int \left| \hat{F}_{\xi}(\mathbf{k}) \right|^{2} d^{3}k. \tag{6.6}$$

More generally, it also follows that the inner product represented by these state vectors can be easily evaluated in **k**-space.

$$\langle \Psi_1 | \Psi_2 \rangle = \int \hat{F}_{\xi 1}{}^{*}(\mathbf{k}) \hat{F}_{\xi 2}(\mathbf{k}) d^3 k \tag{6.7}$$

As a result of the Gaussian spreading function, there will be some overlap between position eigenstates, so that they are no longer strictly orthogonal.

However, by introducing this framework one can evaluate the inner products of states in **k**-space. This **k**-space is clearly the usual momentum space as the Fourier transform expands the wave function in the basis of the eigenstates of momentum (6.4). The crucial expressions here are the inner product (6.7) and the definition of the Fourier transform on which the ability to work in **k**-space is based. In this manner, recalling our caveat for the use of Dirac delta functions, the position of single particles in nonrelativistic quantum mechanics can be treated with relative clarity and consistency.

Relativistic Localization

When one looks to relativistic quantum mechanics, then, this approach must be modified. To begin with, the inner product (6.7) above is not relativistically invariant. In the relativistic case, one would like to express all of this in terms that are equally valid across reference frames. Specifically, one would like to find an invariant inner product and, using an analogous approach to the nonrelativistic case, an invariant Fourier transform.

Relativistic Wave-Packets and the Newton-Wigner State

In making this transition, one may develop a relativistic formalism for localization that makes use of the so-called Newton-Wigner state. To begin, note that in place of the similar expression above (6.4), the position eigenstate should look something like:

$$\Psi(\mathbf{x}, t) = (2\pi)^{-1/2} \int \frac{F(\mathbf{k})}{w(\mathbf{k})} e^{i(\mathbf{k}\cdot\mathbf{x} - w(\mathbf{k})t)} d^3 k \tag{6.8}$$

where

$$w(\mathbf{k}) \equiv \sqrt[+]{1 + \mathbf{k}^2}, \tag{6.9}$$

which comes from the requirement that a free particle be on the mass shell. This constraint may be derived as follows from the energy-momentum four-vector.

$$\begin{pmatrix} E/c \\ p_x \\ p_y \\ p_z \end{pmatrix}, \text{ using the Lorentz norm then, } \frac{E^2}{c^2} - \mathbf{p}^2 = m^2c^2$$

$$\Rightarrow E^2 = \mathbf{p}^2c^2 + m^2c^4$$

$$\Rightarrow E = \sqrt{\mathbf{p}^2 + m^2}, \text{ (if we take } c = 1)$$

Choosing units such that mass, m, is also equal to one, and reverting to $\mathbf{k}$ in place of $\mathbf{p}$ for momentum, E is given by $w(\mathbf{k})$ as in (6.9).

In addition, one may note that in expression (6.8) for the position eigenstate of a scalar spinless particle, $\mathrm{F}(\mathbf{k})$ is a *scalar* function of $\mathbf{k}$; that is to say that $\mathrm{F}'(\mathbf{k}') = \mathrm{F}(\mathbf{k})$ where the primed terms have been Lorentz-boosted into a moving frame. This is the only case considered in this chapter. A crucial feature of this expression is that the term $\frac{d^3k}{w(\mathbf{k})}$ is an invariant volume measure. If the integral is relativistically invariant then $\mathrm{F}(\mathbf{k})$ in (6.8) can be seen as a covariant Fourier transform of $\Psi(\mathbf{x},t)$. Henceforth, the covariant Fourier transform will be denoted with a double hat, i.e., $\hat{\hat{F}}(\mathbf{k})$ will take the place of $\mathrm{F}(\mathbf{k})$.

Similarly, one can define a covariant inner product,

$$\langle \Psi_1 | \Psi_2 \rangle = \int \hat{\hat{F}}_{\xi 1}{}^{*}(\mathbf{k}) \hat{\hat{F}}_{\xi 2}(\mathbf{k}) \frac{d^3k}{w(\mathbf{k})}, \tag{6.10}$$

Setting $t = 0$ in (6.8) produces:

$$\Psi(\mathbf{x},0) = (2\pi)^{-\frac{1}{2}} \int e^{i\mathbf{k}\cdot\mathbf{x}} \hat{\hat{F}}(\mathbf{k}) \frac{d^3k}{w(\mathbf{k})}. \tag{6.11}$$

From these two expressions one can now derive an acceptable invariant position eigenstate. As in the nonrelativistic case, the covariant inner product should provide an orthogonality condition.

$$\begin{aligned} \langle \Psi_\xi | \Psi_{\xi'} \rangle &= \int \hat{\hat{F}}_{\xi}{}^{*}(\mathbf{k}) \hat{\hat{F}}_{\xi'}(\mathbf{k}) \frac{d^3k}{w(\mathbf{k})} \\ &= \delta^3(\xi - \xi') = \langle \phi_\xi | \phi_{\xi'} \rangle \end{aligned} \tag{6.12}$$

Using Parseval's theorem again one can see that the inner product of the nonrelativistic position eigenstate $\phi_\xi(\mathbf{x})=\delta^3(\mathbf{x}-\xi)$ can be written in terms of its Fourier transform.

$$\langle \phi_\xi | \phi_{\xi'} \rangle = \int \hat{\phi}^*(\mathbf{k})\hat{\phi}_{\xi'}(\mathbf{k})d^3k \tag{6.13}$$

where

$$\hat{\phi}_\xi(\mathbf{k}) = (2\pi)^{-1/2}\int_{-\infty}^{\infty}\delta^3(\mathbf{x}-\xi)e^{-i\mathbf{k}\cdot\mathbf{x}}d^3x = (2\pi)^{-1/2}e^{-i\mathbf{k}\cdot\xi}. \tag{6.14}$$

This implies that

$$\hat{\hat{F}}_\xi(\mathbf{k}) = \sqrt{w(\mathbf{k})}\cdot\hat{\phi}_\xi(\mathbf{k}) = \sqrt{w(\mathbf{k})}(2\pi)^{-1/2}e^{-i\mathbf{k}\cdot\xi}. \tag{6.15}$$

Inserting this into the expression for the covariant Fourier transform above results in the well-known Newton-Wigner position eigenstate:

$$\Psi_\xi^{NW}(\mathbf{x}) = (2\pi)^{-3}\int e^{i\mathbf{k}\cdot(\mathbf{x}-\xi)}\frac{d^3k}{w(\mathbf{k})^{1/2}}. \tag{6.16}$$

It is clear that the Newton-Wigner state differs from the nonrelativistic position eigenstate, represented here using a Dirac delta function. Using Basset's formula, one can see that the wave function is spread out in **x** and looks like a modified Hankel function of the first kind, $K_{5/4}(|\mathbf{x}-\xi|)$ in Basset's notation.[20] As is well known, the K-functions behave asymptotically such that they fall off exponentially with the magnitude of the argument.[21] This being the case, the concept of localization must be altered in that each Newton-Wigner state picks out a single value of ξ about which the Hankel function is centered. This may seem an unsatisfactory interpretation for a position eigenstate because it lacks the sharpness of localization implied by a Dirac delta function in **x**, but this is in a sense a trade-off in order to gain covariant eigenstates and a covariant inner product.

In fact one could have chosen a different function of **k**, which would result in a sharply localized function in **x**; following the same approach above, one could set

$$\hat{\hat{F}}_\xi(\mathbf{k}) = w(\mathbf{k})\cdot\hat{\phi}_\xi(\mathbf{k}) = w(\mathbf{k})(2\pi)^{-1/2}e^{-i\mathbf{k}\cdot\xi}, \tag{6.17}$$

and, inserting this into the expression for the covariant Fourier transform, this becomes:

$$\Psi_\xi(\mathbf{x}) = (2\pi)^{-3} \int e^{i\mathbf{k}\cdot(\mathbf{x}-\boldsymbol{\xi})} d^3k = \delta^3(\mathbf{x} - \boldsymbol{\xi}) \tag{6.18}$$

which is a delta function in three spatial dimensions. One can see the price of this trade-off by looking at the inner product this provides:

$$\langle \Psi_\xi | \Psi_{\xi'} \rangle = \int \hat{\hat{F}}_{\xi'}(\mathbf{k}) \frac{d^3k}{w(\mathbf{k})}$$
$$= (2\pi)^{-3} \int e^{ik\cdot(\boldsymbol{\xi}-\boldsymbol{\xi}')} w(\mathbf{k}) d^3k \neq \delta^3(\boldsymbol{\xi} - \boldsymbol{\xi}').$$

This does not display the orthogonality condition needed. Thus, it is preferable to use Newton-Wigner states as the position eigenstates in spite of the unusual interpretation of localization they imply.

Next, one would like to express a general state, eventually a compact particle in a box, as a sum over the basis provided by Newton-Wigner eigenstates of position.

$$|\Psi\rangle = \int \langle \Psi_\xi^{NW} | \Psi \rangle |\Psi_\xi^{NW}\rangle d^3\xi$$

where $|\Psi_\xi^{NW}\rangle$ is the Newton-Wigner state, so that

$$|\Psi\rangle = \int G(\boldsymbol{\xi}) |\Psi_\xi^{NW}\rangle d^3\xi.$$

Now $G(\boldsymbol{\xi})$ represents the probability density amplitude, in a general state $|\Psi\rangle$, of finding a particle at position $\boldsymbol{\xi}$, and can be rewritten using the expression for the covariant inner product (6.10).

$$G(\boldsymbol{\xi}) = (2\pi)^{-1/2} \int \left(e^{ik\cdot\boldsymbol{\xi}} \sqrt{w(\mathbf{k})}\right) \hat{\Psi}(\mathbf{k}) \frac{d^3k}{w(\mathbf{k})}. \tag{6.19}$$

The probability density of finding a particle at position $\boldsymbol{\xi}$ is then expressed by $|G(\boldsymbol{\xi})|^2$. The covariant Fourier transform of the Newton-Wigner state is already known from its derivation (6.15); substituting this into (6.19), this becomes:

$$G(\boldsymbol{\xi}) = (2\pi)^{-1/2} \int \left(e^{ik\cdot\boldsymbol{\xi}} \sqrt{w(\mathbf{k})}\right) \hat{\Psi}(\mathbf{k}) \frac{d^3k}{w(\mathbf{k})}. \tag{6.20}$$

From the covariant Fourier transform (6.11), one concludes that

$$\Psi(\mathbf{x},0) = (2\pi)^{-\frac{1}{2}} \int e^{i\mathbf{k}\cdot\mathbf{x}} \hat{\hat{\Psi}}(\mathbf{k}) \frac{d^3k}{w(\mathbf{k})}, \tag{6.21}$$

which is then simply inverted using the inverse (noncovariant) Fourier transform such that

$$\frac{\hat{\hat{\Psi}}(\mathbf{k})}{w(\mathbf{k})} = (2\pi)^{-\frac{1}{2}} \int e^{-i\mathbf{k}\cdot\mathbf{x}} \Psi(\mathbf{x}) d^3x. \tag{6.22}$$

Substituting (6.22) into (6.20), one gets another expression for G(ξ).[22]

$$G(\xi) = (2\pi)^{-3} \int\int \left(e^{i\mathbf{k}\cdot(\xi-\mathbf{x})}\sqrt{w(\mathbf{k})}\right)\Psi(\mathbf{x})d^3k\,d^3x \tag{6.23}$$

Although this expression has been derived by taking $\Psi(\mathbf{x})$ as the state at $t = 0$, the same formula applies for the state at an arbitrary time t. One may see this clearly by recalling that one must now take the inner product of Ψ with the Newton-Wigner states that are localized *at time t*.

Using the two expressions derived above, (6.20) and (6.23), and recalling that G(ξ) is the probability density amplitude in a general state $|\Psi\rangle$ for finding a particle at position ξ, one may now note some of the difficulties that arise with the formulation of localized states in relativistic quantum mechanics.

Problems with Relativistic Localization

The difficulties with relativistic localization are evident in the Newton-Wigner position state. The root problem is once again that of nonlocality, which one would have hoped would be less of a problem within an explicitly relativistic formulation of quantum theory. This root problem manifests itself in what may be called the "Jericho effect," the failure of the localized state to remain invariant under Lorentz transformations.

Non-Invariance: The Jericho Effect

First consider what happens to the Newton-Wigner state $|\Psi^{NW}\rangle$ localized at the point $\xi = 0$, when it is boosted into a moving frame. Starting from the stationary frame, one notes from (6.15) that

$$\hat{\hat{\Psi}}^{NW}(\mathbf{k}) = \sqrt{w(\mathbf{k})}(2\pi)^{-1/2}e^{-i\mathbf{k}\cdot\xi}$$
$$= \sqrt{w(\mathbf{k})}(2\pi)^{-1/2}, \text{ for } \xi = 0. \tag{6.24}$$

where the double hat denotes the covariant Fourier transform. Substituting (6.24) into (6.20), one can see that the amplitude G(ξ) before the boost is:

$$G(\xi) = (2\pi)^{-3}\int \left(e^{i\mathbf{k}\cdot\xi}\sqrt{w(\mathbf{k})}\right)\sqrt{w(\mathbf{k})}\frac{d^3k}{w(\mathbf{k})}$$
$$\Rightarrow G(\xi) = (2\pi)^{-3}\int e^{i\mathbf{k}\cdot\xi}\hat{\Psi}(\mathbf{k})d^3k = \delta^3(\xi) \tag{6.25}$$

as might be expected. As previously noted, these scalar functions in momentum space provide a simple way to boost the state into a moving frame for a spinless particle. Thus, under a Lorentz transformation, equation (6.24) becomes

$$\hat{\hat{\Psi}}^{NW'}(\mathbf{k}) = \sqrt{w'(\mathbf{k})}(2\pi)^{-1/2}. \tag{6.26}$$

Applying this to (6.20), one can boost the state into a moving (primed) frame so that

$$G(\xi') = (2\pi)^{-3}\int e^{i\mathbf{k}\cdot\xi}\sqrt{w'(\mathbf{k})}\frac{d^3k}{\sqrt{w(\mathbf{k})}} \neq \delta^3(\xi'). \tag{6.27}$$

The inequality here is due to the fact that the $w(\mathbf{k})$ terms do not cancel one another as they did before the boost (6.25).

To examine the situation more explicitly, one may look at the case of a single spatial dimension. Following from (6.16), the Newton-Wigner state localized at $\xi = 0$ is given by:

$$\Psi_0^{NW}(x) = (2\pi)^{-1}\int_{-\infty}^{\infty}\frac{e^{ikx}}{w(k)^{1/2}}dk \propto x^{1/4}K_{1/4}(x). \tag{6.28}$$

Now in order to arrive at the amplitude for finding the particle at position ξ' in a Lorentz-boosted frame, the following integral must be evaluated:

$$G(\xi') = (2\pi)^{-1} \int_{-\infty}^{\infty} e^{ik\xi'} \sqrt{\frac{w'(k)}{w(k)}} dk. \tag{6.29}$$

This follows from expression (6.27) given above in three spatial dimensions. To this end, one can approximate $w'(k)$ for small boosted frame velocities, v. One may first of all note that, as scalar functions, which follows from (6.9). In addition,

$$k = (1 - v^2)^{½}(k' + vw') = (1 - v^2)^{½}\left(k' + v\sqrt{1 + k'^2}\right),$$

which leads to the approximation that

$$k^2 \cong k'^2 + 2vk'\sqrt{1 + k'^2}$$

to the order of velocity, v. Hence one finds that

$$w(k) \cong w(k')\sqrt{1 + \frac{2vk'}{\sqrt{1 + k'^2}}} \cong w(k')\left(1 + \frac{vk'}{w(k')}\right).$$

This leads to the conclusion that

$$w'(k') = w(k) \cong w(k') + vk', \tag{6.30}$$

or, replacing k' by k,

$$w'(k) \cong w(k) + vk. \tag{6.31}$$

Finally, one may claim:

$$\sqrt{\frac{w'(k)}{w(k)}} \cong 1 + \frac{1}{2} v \frac{k}{\sqrt{1 + k^2}}. \tag{6.32}$$

Applying this approximation to (6.29) leads to the probability density amplitude (exchanging ξ' by ξ):

$$\begin{aligned} G(\xi) &\cong \delta(\xi) + v(4\pi)^{-1} \int_{-\infty}^{\infty} e^{ik\xi} \frac{k}{\sqrt{1 + k^2}} dk \\ &= \delta(\xi) + \frac{iv}{2\pi} K_1(\xi). \end{aligned} \tag{6.33}$$

This expression is analogous to (6.27), which gives the same amplitude in three spatial dimensions. In this case, however, it is clear that the second term of (6.33), which is linear in v, displays the first-order departure of $G(\xi)$ from the pure delta function $\delta(\xi)$.

All of this is to confirm that in general the probability density amplitude $G(\xi)$ is frame dependent, a clear source of difficulty with Newton-Wigner localization. The surprising implication of this is that it is not possible to describe a physical system or particle as objectively localized from the point of view of all reference frames. In perspectival invariantist terms, the Newton-Wigner state is a representational structure that fails to be objective *(Obj_P)* by failing to remain invariant under Lorentz transformations, members of the Poincaré group. Thus the representational relation between this localized relativistic quantum state and a single particle fails to be objective, in the sense of perspectival invariantism.

The noninvariance of the Newton-Wigner state under Lorentz transformation, the Jericho effect, alludes to the biblical story in which the walls of the ancient city came down as the men of Israel marched around them.[23] Similarly, the probability distribution of a localized particle here goes from being sharply bounded to having infinite tails when the observer is boosted into a moving frame. One reason for attaching this name to this consequence of relativistic quantum theory is to emphasize the counterintuitive implications regarding localization. In other words, it seems that behavior of particles in this manner ought to belong to the miraculous.

One might respond by suggesting that this case is fundamentally ill-suited to the Jericho comparison: No walls have been explicitly considered, Newton-Wigner states being free particle solutions of the Klein-Gordon equation. Furthermore, it might be thought that by explicit introduction of infinite potential barriers, a more familiar, relativistically invariant localization might be achieved. In fact, neither suggestion is correct. These responses emphasize further the surprising nature of the Jericho effect, as it is in precisely this case, with infinite potential barriers, that localization is frame dependent.

To see this more clearly, one must remember that a position probability distribution that goes to a zero value outside of a finite region is a way of describing the effect of infinite potential barriers. With this in mind, it is clear that the walls of Jericho, are indeed accounted for in the treatment above. In fact, one might think of a single Newton-Wigner state as that particle in a box for which the dimensions of the box have been reduced to a single point. This state might be prepared with two pistons pressing on the sides of the box until the probability amplitude of finding the particle is a delta function at a single point. These pistons could be represented by an infinite potential well. Regardless of the way in which they are prepared, the Newton-Wigner states provide a perfectly valid way of thinking about localized states. Furthermore, this is the same spirit in which Hegerfeldt considers localization in the theorem that bears his name. As it is this theorem that is being elucidated here, it seems appropriate to follow this line.[24]

Time Evolution in a Single Frame: Hegerfeldt Nonlocality

In addition to the noncovariance of $G(\xi)$, another difficulty arises from the time evolution of Newton-Wigner localized states in a single frame. Proceeding as above and reintroducing time dependence, one can see from (6.16) that

$$\Psi(\mathbf{x},t) = (2\pi)^{-3} \int e^{i(\mathbf{k}\cdot\xi - w(\mathbf{k})t)} \tag{6.34}$$

for a state localized at the origin at time $t = 0$. The probability amplitude at time t is given by:

$$G(\xi) = (2\pi)^{-3} \int e^{i(\mathbf{k}\cdot\xi - w(\mathbf{k})t)} d^3k \tag{6.35}$$

where $w(\mathbf{k}) = \sqrt{m^2 + \mathbf{k}^2}$. This integral has branch points at $|\mathbf{k}| = \pm im$, so that $G(\xi)$ cannot vanish even at very large distances $|\xi|$, as discussed in detail by Hegerfeldt.[25] This implies that if state $|\Psi\rangle$ is localized for time $t = 0$, that at any later time $G(\xi)$ will be nonzero everywhere. Since $|G(\xi)|^2$ represents the probability density of finding a particle at a position ξ, this entails an apparent violation of Einstein locality; the particle has a nonzero probability of traveling faster than c.

The problem of Newton-Wigner time evolution is essentially equivalent to that of Hegerfeldt-type nonlocality. Hegerfeldt's theorem deals with just this problem in the time evolution of Newton-Wigner localized states. The evaluation of the Hegerfeldt integral (6.35) as nonzero everywhere for arbitrarily small t implies the possibility of a nonlocal phenomenon.

Hegerfeldt's Integral

Although Hegerfeldt himself comes to this conclusion by evaluating this expression as a complex contour integral, it is instructive to solve it directly. The case of one spatial dimension will be considered and two methods are attempted for small time displacements and large distances, ξ, from the position where the state is initially localized.

The motivation behind going through these calculations is to demonstrate the role of the Jericho effect in the commonly accepted Hegerfeldt nonlocality result, in a way that brings out the physical argument not apparent in the method presented above. Lorentz transformations are generally thought of as kinematical in special relativity whereas the time evolution of the state vector is by definition dynamical. The connection between the Jericho effect and the Hegerfeldt result is then a demonstration of the meshing of kinematics and dynamics that takes place in relativistic quantum mechanics. Alternatively, one might say that Lorentz boosting in relativistic quantum theory is not kinematical at all but clearly dynamical, which amounts to a statement supporting the same conclusion.

Direct Solution for Small Time Displacement

Starting with the integral that Hegerfeldt considers, (6.35), one may focus on the regime where time displacement is very small. So, for $t = \varepsilon \ll 1$, and simplifying the integral to one spatial dimension, it becomes

$$G(\xi) = (2\pi)^{-1} \int_{-\infty}^{\infty} e^{ik\xi}(1 - iw(k)\varepsilon)dk$$

$$\Rightarrow G(\xi) = (2\pi)^{-1} \int_{-\infty}^{\infty} e^{ik\xi}dk - \frac{i\varepsilon}{\pi} \int_{-\infty}^{\infty} \sqrt{1+k^2}\cos(k|\xi|)dk \qquad (6.36)$$

$$\Rightarrow G(\xi) = \delta(\xi) - \frac{i\varepsilon}{\pi} \cdot I.$$

The approximation for small t is appropriate as one is interested here in the amplitude $G(\xi)$ in just this regime where nonlocal effects appear. Further simplifying by solving the integral I in (6.36) produces:

$$I = \int_{0}^{\infty} \sqrt{1+k^2}\cos(k|\xi|)dk = \left(1 - \frac{d^2}{d|\xi|^2}\right)\int_0^\infty \frac{\cos(k|\xi|)}{(1+k^2)^{1/2}}dk. \quad (6.37)$$

Using Basset's formula, this simplifies to

$$I = \left(1 - \frac{d^2}{d|\xi|^2}\right) K_0(|\xi|) - K_0''(|\xi|).$$

Now using the following recurrence relations,[26]

$$\begin{aligned} K_0' &= -K_1 \\ K_1' &= -\tfrac{1}{2}(K_0 + K_2) \\ \tfrac{-2}{\xi} K_1 &= \tfrac{1}{2}(K_0 - K_2) \end{aligned} \tag{6.38}$$

leads to the conclusion that:

$$\mathrm{I} = K_0 - K_0'' = K_0 + K_1' = \tfrac{1}{2}(K_0 - K_2) = \tfrac{-2}{|\xi|} K_1. \tag{6.39}$$

In expression (6.39), K_1 has an asymptotic form such that $I \cong \frac{-2}{|\xi|}\sqrt{\frac{\pi}{2|\xi|}}e^{-|\xi|}$; substituting back into the expression for the $G(\xi)$ results in:

$$G(\xi) \cong \delta(\xi) + i\varepsilon\sqrt{\tfrac{2}{\pi}} \cdot \frac{e^{-|\xi|}}{|\xi|^{3/2}} \tag{6.40}$$

The delta function naturally goes to zero for large values of ξ, so that asymptotically

$$G(\xi) \cong i\varepsilon\sqrt{\tfrac{2}{\pi}} \cdot \frac{e^{-|\xi|}}{|\xi|^{3/2}} \text{ and } |G(\xi)|^2 = \tfrac{2}{\pi} \cdot \frac{e^{-2|\xi|}\varepsilon^2}{|\xi^3|}. \tag{6.41}$$

The equations in (6.40) and (6.41) represent the exact solutions for the Hegerfeldt integral for sufficiently small t; it is worth noting that for $t = 0$, $G(\xi)$ is simply a delta function at $\xi = 0$, as one should expect, and that the amplitude has infinite tails that are generally speaking exponential. All that is needed for a nonlocal effect is a nonzero probability for large distances and small t, but one may go further and show that these exponential tails can be understood as arising from the Jericho effect. That is to say they can be seen as brought about by contributions to $G(\xi)$ from hyperplanes associated with boosted frames.

Approximate Method Using Hyperplane Geometry

In order to see this one can solve for $G(\xi)$, the probability amplitude for finding a particle at a position ξ, in a way that makes explicit use of the geometry of hyperplanes in Minkowski spacetime. As before, the desire is to solve for $G(\xi)$ at some time t arbitrarily close to $t_o = 0$, the instant that the Newton-Wigner localized wavepacket has begun to spread; that is to say the moment that the particle has been let out of the box. Recalling one of our expressions (6.23) for $G(\xi)$, adapted to one spatial dimension, produces:

$$G(\xi) = (2\pi)^{-1} \int_{-\infty}^{\infty}\int_{-\infty}^{\infty} \left(e^{ik(\xi - x)}\sqrt{w(k)}\right)\Psi(x)dkdx. \tag{6.42}$$

Reverting to the remarks made with reference to (6.23), one may take for $\Psi(x)$ the spatial dependence of $\Psi(x,t)$ for a small time interval ε after the particle escapes from the box. As $\Psi(x,t)$ is a scalar function, it can be expressed in terms of a moving frame as $\Psi'(x',t')$. In particular, one may choose the intersection of the two hyperplanes and note that for a scalar particle

$$\Psi(x,t) = \Psi'(x'(x,t),0) \tag{6.43}$$

Variables x, x', t, and t' are all related by the familiar Lorentz transformations:

$$\begin{aligned} x' &= \gamma(x - vt) \\ x &= \gamma(x' + vt) \\ t' &= \gamma\left(t - vx / c^2\right) \\ t &= \gamma\left(t' + vx' / c^2\right) \end{aligned} \tag{6.44}$$

where $\gamma = (1 - v^2/c^2)^{-1/2}$ as usual. Now, $\Psi'(x',0)$, can be expressed using the covariant Fourier transform (6.11) adapted to one spatial dimension, as

$$\Psi'(x',0) = (2\pi)^{-1/2} \int_{-\infty}^{\infty} e^{ik'x'}\,\hat{\hat{F}}_0'(k')\frac{dk'}{w(k')}. \tag{6.45}$$

Recalling that in this case, $\hat{\hat{F}}_0'(k') = (2\pi)^{-1/2}\sqrt{w'(k')}$ (see [6.15]), and substituting (6.43) and (6.45) into (6.42), results in:

$$G(\xi) = (2\pi)^{-2} \int_{-\infty}^{\infty}\int_{-\infty}^{\infty}\int_{-\infty}^{\infty} e^{ik'x'(x,t)}\,\frac{w'(k')^{1/2}\cdot w(k)^{1/2}}{w(k')}e^{ik(\xi - x)}dkdxdk'. \tag{6.46}$$

It should be noted that the method of evaluating $\Psi(x,t)$ expressed in (6.43) implies that every point (x,t) can also be described, in terms of the moving frame, as $(x'(x,t),\ t' = 0)$; this means that for each (x,t) there is a velocity, v, of the moving frame for which (6.43) holds. As v is in general bounded by c such that $-c < v < c$, then only points (x,t) up to the light-cone can be expressed as $(x'(x,t),0)$. Since one is interested here in small intervals of time ε where $t = t_o + \varepsilon$, by moving to arbitrarily small ε the boundaries on the validity of the condition that $(x,t = \varepsilon) = (x'(x,t),\, t' = 0)$ imposed by the light-cone also become arbitrarily small.

It is helpful to depict this in 1 + 1 dimensional Minkowski spacetime. See Figure 6.1. Spacetime point P has coordinates $(x',0)$ in the moving frame and (x,ε) in the stationary frame. Here one is evaluating $\Psi(x,\varepsilon)$ at P as identical with $\Psi'(x',0)$ along the inclined hyperplane. One is thus employing the Jericho effect to pick up larger amplitudes in the boosted frame than would occur if the particle were Newton-Wigner localized in the boosted frame. This failure to be localized in a boosted frame is of course the essence of the Jericho effect.

Adopting this approach, then the expression for $G(\xi)$ has the final form, for a small time interval $t = \varepsilon$,

$$G(\xi) = \delta(\xi) + i\varepsilon f(\xi)e^{-|\xi|} \tag{6.47}$$

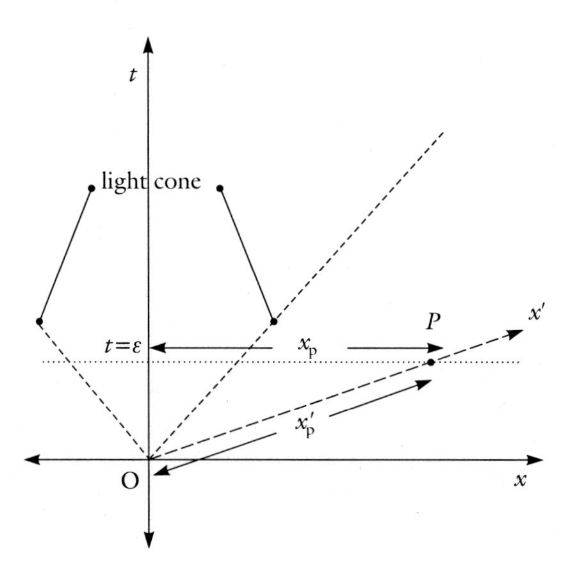

Figure 6.1. In 1 + 1 dimensional spacetime, point P marks the intersection of the two hyperplanes of simultaneity associated with resting and boosted reference frames (depicted by a dotted line and a dashed arrow, respectively).

where $f(\xi)$ does not interfere with the exponential asymptotic behavior for large ξ.[27]

The form of this second result, (6.47), is basically that expected from the direct solution of the Hegerfeldt integral for small t, (6.40) above. This method, while not as rigorous, suggests that the nonlocal spreading of the wave-packet can be seen as due to the Jericho effect. In other words, the two difficulties with Newton-Wigner localization, Hegerfeldt nonlocality (time evolution) and the Jericho effect (the noncovariance of localization on shifted hyperplanes), are essentially related phenomena.[28]

Localization and Perspectival Invariance

These examples should serve to illustrate that representing a localized quantum state within a relativistic context is, to say the least, problematic. One may attempt to represent such a structure using the Newton-Wigner position state. However, this leads to the Jericho effect (or equivalently Hegerfeldt nonlocality).

These difficulties are historically significant in that they have led in part to the rejection of the relativistic quantum mechanics of particles in favor of quantum field theory. As previously noted, David Malament has argued that these sorts of difficulties with localization rule out the possibility of representing localized particles using relativistic quantum mechanics.[29] Indeed, in quantum field theory one finds that the notion of a single localized particle is effectively abandoned in favor of thinking in terms of charge density. Thus relativistic quantum field theory avoids the difficulties encountered above in trying to represent a localized state in relativistic quantum mechanics.

There is, however, another option. This is to retain the formalism of relativistic quantum mechanics within the so-called hyperplane-dependent interpretation championed by Gordon Fleming.[30] Under this scheme localization of a particle is still possible by parameterizing the state to individual inertial hyperplanes of simultaneity. The hyperplane-dependent approach, as should be evident from the name, takes as its starting point a view of special relativity that makes use of the concept of the relativity of simultaneity. As previously discussed, this is the view that each inertial frame in special relativity has its own distinct hyperplane of simultaneity that may be defined in terms of Minkowski orthogonality to an inertial worldline.

Following the hyperplane-dependent approach to localization of a single particle state, one would claim that the Newton-Wigner state is in reality a perfectly objective *(Obj_P)* means of representing a localized particle within the limited context of a single rest frame. However, one should not expect the same state to remain localized in any other frame of reference, or equivalently on any other hyperplane of simultaneity. The fact that such a state does not remain localized under changes to different frames once again reflects the essence of the Jericho effect presented above. Following the same approach in every rest frame, one may construct an infinite number of localized states associated with each hyperplane of simultaneity. Thus position may no longer be characterized simply by spatial parameters, but also by the parameters that refer to different hyperplanes. As demonstrated in Chapter 4, the choice of which hyperplane of simultaneity to use in a given representation is absolutely conventional so that these parameters may be thought of in terms of either the velocity, v, of moving reference frames or the Reichenbach standard of synchrony, ε (see equation [4.4]). In choosing to adopt this solution to the problem of localization, one ends up with a workable localized position state, but one that differs considerably from traditional notions of position in its hyperplane dependence.

The hyperplane-dependent approach may also be seen as a means of recovering objectivity *(Obj_P)* in representation, making use of the same relational strategy employed in response to Weyl in the treatment of a vertical vector in Chapter 3. Recalling this example, a relational strategy was employed by altering the notion of vertical to vertical at a given position, which was (trivially) invariant under spatial translations. This strategy succeeded in making the vector objective but is also equivalent to removing spatial translations from the relevant invariance criterion, the Euclidean group. Hyperplane dependence, as a response to the problem of localization in relativistic quantum mechanics, is just such a relational strategy. The notion of hyperplane-dependent localization is fully objective *(Obj_P)*, but the restriction of localization to individual hyperplanes effectively removes Lorentz transformations from the relevant invariance criterion, in this case the Poincaré group.

These two options, quantum field theory and hyperplane dependence, once again illustrate a fundamental feature of perspectival invariantism. This is the inverse relationship between invariant features of a representation and the size of the group of transformations that comprise its invariance criterion. In relativistic quantum mechanics, a localized quantum state fails to be Lorentz invariant. The two possible responses involve either the choice to adopt quantum field theory, on the one hand, or a hyperplane-dependent notion of localization, on the other. According to the former, the invariance criterion remains the Poincaré group, but one must abandon the concept of sharp particle localization. According to the latter, one retains sharp localization, but the invariance criterion effectively loses the Lorentz transformations as members. Thus with a larger group one has less invariant features than with the effectively smaller group associated with hyperplane dependence.

Conclusion

The simplest and yet most fundamental intuition that runs through this book is that the physical sciences, conventional though they are in many ways, may nevertheless provide one with objective representations of reality. Here, this general point has been illustrated in the context of the debate over the philosophical significance of symmetry in modern science. It is hoped that this will contribute to weakening the hold of the apparent tension often assumed to exist between objectivity and conventionality in the physical sciences. One ought to see no paradox in the fact that scientific models represent objective effects, like the different age of Langevin's space traveler, as well as merely conventional ones, like the time dilation of special relativistic kinematics. In conclusion, consider again how a new appraisal of symmetry supports this conclusion.

To begin with, a view of representation was proposed that explicitly includes the human subjects between whom representation actually takes place. This view takes account of both the social mediation of information and the mediation of formal relations between structures. In considering the formal and social dimensions of scientific representation, human subjects must often resolve ambiguity through conventional choice. Thus, scientific representation relies on three basic relations: isomorphic maps, the representational relation, and the information relation. While isomorphisms relate the structural features of elements of a formal chain of media, the representational relation, on the other hand, is a conceptual relationship that holds between two structures such that one is considered to be a model of the other; thus this relation links a model with an original or a token with a type. The information relation, by contrast, links two people, members of a social chain of media.

One formal feature of the models physicists use to represent the world is captured by their symmetries, where a symmetry transformation, or automorphism, maps a structure back into itself. These transformations may also be factorized in terms of maps between different isomorphic structures. Both automorphisms and isomorphisms are used to handle the representational structures that together comprise the formal chain of media. This formal chain of media is a collection of structures, each of which retain the invariant features that the model and original must share, and the different members of this chain are linked to one another through strict or partial isomorphic maps.

Given this background, one is in a better position to consider the striking claims of invariantism. Hermann Weyl famously expressed the view that "objectivity means invariance with respect to the group of automorphisms,"[1] which may be taken as its slogan. Upon close consideration, one may indeed reject the general invariantist claim that links the heuristic and unifying roles of symmetry to scientific objectivity.

Having arguably rejected invariantism in its full form, one is nevertheless able to construct a limited version of the position: perspectival invariantism. However, even within representations that are objective in this sense, ambiguities in general present themselves that must be resolved through various kinds of conventional choice. In particular, this form of objectivity, termed perspectival objectivity *(Obj_P)*, requires the features of a given model to be invariant under a group of automorphisms that meets certain criteria. Namely, the symmetries must admit of interpretation as changes in idealized perspective and must have proven themselves to be generalizable and heuristically fruitful. When these criteria are met, a group of automorphisms may then be considered as an invariance criterion for the objectivity *(Obj_P)* of a given model.

This approach leaves one with the question of how this form of objectivity relates to more traditional views on the meaning of the term. Objectivity is, curiously, both much debated and neglected within the philosophy of science. It is much debated in the sense that it is often highly contested, yet its relative neglect is evident in that no single definition of the term dominates, even as a foil for critique; as one has for instance in the "justified true belief" account of knowledge. In order to account for how perspectival objectivity compares to the numerous other views on the subject, two major senses of the term "objectivity" were discussed: objectivity as multi-subjectivity and objectivity as ontological subject independence.

Perspectival objectivity falls within the category of the former, multi-subjectivity, since it is based on the intuition that objective facts are those which appear the same from different perspectives and are thus in principle agreed upon by subjects occupying those points of view. Traditionally, one problem with multi-subjective views of objectivity is that many definitions, for example the sociological definitions of objectivity, have taken objectivity to be constrained only by social interests and networks. Here perspectival objectivity is multi-subjective in its insistence that subjects agree on certain things, but is also heavily constrained by the formal aspects of the representation. Specifically, the idealized perspectives under consideration here are a function of the structure of the models under consideration, not primarily of the social context of the subjects. Hence this is not an "anything goes" notion of objectivity; it is, however, a notion that leaves considerable room for conventional choices.

The conventional choices that present themselves in scientific representation may be grouped into several categories: trivial, relational, and absolute. Trivial conventional ambiguity is that which exists in, for instance, the semantic decision over whether to refer to a certain entity as a "dog" or a "car." Clearly either choice, applied consistently, should be acceptable, and will not lead to further ambiguity. Relational conventions, like Reichenbach's coordinative definitions, relate representational structures to one another in a way that allows one to carry out the manipulations and calculations needed in modern physics. Absolute conventions are the most problematic. As in the trivial case there seems to be no fact-of-the-matter in choosing one of these conventions over another, and yet they do seem to provide certain constraints on the way one represents reality; the classic example of this is conventionality of simultaneity. All three of these forms of conventionalism, and most notably the last, are still evident within objective *(Obj$_P$)* representations in physics.

The three cases that were presented in Chapters 4–6 each illustrate certain of the key issues raised above. In order to see this, consider once more the significance of the treatment given to the conventionality of simultaneity, the twin paradox, and localization in relativistic quantum theory.

The first case addresses the debate over distant synchrony in special relativistic kinematics. This debate is longstanding, but is as yet not entirely resolved. It is a good illustration of perspectival invariantism since the relevant symmetries may be easily interpreted as perspectives, and these symmetries have proven to be generalizable and immensely heuristically successful. This case also provides a new insight into the philosophical point at issue between proponents of the conventionality of simultaneity, on the one hand, and the relativity of simultaneity, on the other. In fact, by adopting a perspectival invariantist point of view, the difference between these positions can be understood as a disagreement over what is objective *(Obj*$_P$*)*. For the conventionalist, any standard of synchrony is purely conventional, while for the relativist, Einstein synchrony is objective in a given frame. This relativized form of objectivity, however, is purchased by effectively excluding the Lorentz transformations from the group of automorphisms that comprise the relevant invariance criterion. Thus, the perspectival invariantist approach to this controversy would indicate that invariance has at least as much to do with convention as it does with objectivity, especially since the ability to claim that simultaneity is objective (in a given frame) depends on a conventional choice as to which symmetries to include within the invariance criterion.

The case of the twin paradox illustrates how invariance can serve to pick out objective *(Obj*$_P$*)* scientific representations. The twin paradox, in its minimal standard version, is used to represent the consequences of special relativistic kinematics. It was show that in reality it depends on the representational relation between elapsed time on a moving clock and proper time calculated along its path. In numerous other versions of the twin paradox, explanations are put forward for the different ages of the twins that are based not on this representational relation, but on one simultaneity convention or another. These versions may be ruled out as not objective *(Obj*$_P$*)*, and the generalized scheme presented illustrates this point in a striking and novel way.

Furthermore, the structures associated with different standards of synchrony, sets of inertial coordinates, are not invariant under the Poincaré group. But this group of symmetry transformations is the most obvious invariance criterion for models in special relativity theory. Because these structures fail to be invariant, then they may not be considered as objective *(Obj*$_P$*)*. Proper time, on the other hand, is invariant; thus representations that make use of this quantity are. These objective representations of the twin paradox nevertheless contain conventional ambiguities. Most significant, the choice of simultaneity criterion remains absolutely conventional.

Turning to the case of localization in relativistic quantum theory, the objectivity of localization may only be recovered by opting for one of two possibilities. The first is quantum field theory, in which the notion of representing single localized particles is abandoned in favor of charge distributions; in this case the representational relation is changed to that between a single localized particle and a charge distribution. The second option is to adopt a hyperplane-dependent approach to the problem. By changing the notion of localization to localization on a given hyperplane, one may, by taking advantage of a relational strategy, ensure that localized states are Lorentz invariant after all. This strategy, however, effectively removes Lorentz transformations from the relevant group of symmetries, changing the invariance criterion.

The result is that in quantum field theory, on the one hand, and hyperplane dependence, on the other, one may recover objective *(Obj$_P$)* representations by changing the representational relation and/or the alignment criterion, respectively. The choice of which way to go in recovering objectivity is one of convention, and physicists might opt for either one on the basis of a number of criteria, but objectivity *(Obj$_P$)* is not one of them. Thus, to the extent that invariance leads to objectivity, it also introduces ambiguities that must be resolved by convention.

In assessing the new appraisal of symmetry presented in this work, one may point first to the refutation of invariantism as a means of accounting for the philosophical significance of symmetry. Of equal importance is the general point that symmetry has at least as much to do with convention as with objectivity, which is bolstered by case studies of perspectival invariantism. Thus, the novel approach offered here has a negative and a positive dimension; the negative is that invariantism is not an accurate assessment of the situation; the positive is that symmetries in models can illuminate the subtle and significant relationship between convention and objectivity in scientific representation.

At this point, the reader may wonder if the real significance of symmetry might be found in a much more open-ended, even empirical, sense in the generalizability and heuristic power of certain symmetries. From a methodological perspective, that is certainly the case. Symmetry, as formalized in group theory, is a powerful mathematical tool. From a philosophical perspective, however, one is always tempted to wonder if there is more that can be said. In reality, trying to account for the conceptual significance of symmetry in an elegant and concise manner is not unlike the attempt to explain why any bit of mathematics has been so fruitful as a tool for representing the natural world. While intuitively one is aware that this is a deep and important question, in a sense it is too deep, and may in the end simply defy philosophical analysis.

But this sense of open-endedness should not be considered a deficiency in this assessment of symmetry in physics. This book began with a reference to Robert Nozick, and a photo of an eye hidden behind a sunglass lens. It is appropriate to close with another reference to Nozick, who warns, "One form of philosophical activity feels like pushing and shoving things to fit into some perimeter of specified shape . . . All that remains [after such activity] is to publish the photograph as a representation of exactly how things are."[2] Some of what has preceded this is about fitting shapes in the literal sense of isomorphism between a model, *M*, and its idealized original, *O*. However, in the sense of philosophical programs, which Nozick intends, this work has argued that the philosophical significance of symmetry in physics does not fit within the boundaries offered by invariantism. Thus, it is hoped that the reader will not feel that Nozick's critique is applicable here. Moreover, there may be other parameters that are equally useful in thinking about symmetry in physics, especially in regard to its generalizability and heuristic power.

As stated at the outset, the significance of this inquiry will depend on one's initial view of the sciences. For the committed conventionalist, it is hoped that this will confirm that conventional ambiguity is not just a feature of the inexact, less mathematically rigorous sciences, but indeed that it is evident within both of the major strands of modern theoretical physics, relativity theory and quantum mechanics. On the other hand, the conventionalist might be disinclined to accept that a meaningful notion of objectivity in scientific representation can coexist with all this conventional ambiguity. For those who find the objectivity of the physical sciences self-evident, perhaps this will confirm the intuition that scientific representation is objective, and do so in a way that also takes seriously the subject-dependent aspect of science in practice. The general principle of retaining a role for both the objective and the conventional, and understanding the way they relate to one another, cannot but help in the attempt to address the numerous philosophical problems that exist in the interpretation of modern science.

Finally, Nozick goes on to ask of the boundaries established by various philosophical programs, "What does having everything within a parameter *do* for us? Why do we want it so?"[3] In this book, the picture painted and the photograph (literally) published, promise to do at least two things: to serve as a reminder that scientific objectivity is inextricably linked (in principle and in practice) to the humans who seem to care so much about it, and that individual perspectives and their associated conventions are similarly inseparable from their grander cousins, the facts. If there is a core philosophical program here, it has to do with keeping both the formal and social dimensions of scientific representation in the frame. Why do we want it so? Because the solipsistic tendencies of extreme conventionalism are just as limiting as the dehumanizing effect of overly simplistic objectivism in physical science.

Appendix A

Appendix B

Notes

Index

Appendix A

To further convince the skeptic about the implications of the Jericho effect, another way of expressing it is to think in terms of position operators. Since in quantum mechanics it is the position operator that generates an eigenstate of position, this should be equivalent to whatever state-localizing account one gives using potentials. When one considers localizing operators, one finds that localization is again frame dependent, another way of stating the Jericho effect. One first recalls that Newton-Wigner localization is not in **x**, but in **ξ**. This is one of the costs of accepting Newton-Wigner position states which, as previously shown, are spread out in **x**-space and only localized in the variable **ξ**.

To bring out the essential point at issue (and continuing the discussion in one spatial dimension), a comparison may be made between the position operator for the rest frame and one for a tilted hyperplane. The operator that localizes a state in the variable **ξ**, in the rest frame, has the form (see Greiner, 73–74)

$$\hat{\xi} = i\frac{d}{dk} - \frac{ik}{2w(k)^2}. \tag{A.1}$$

The first term is recognizable as the position operator in momentum representation in nonrelativistic quantum mechanics (taking $\hbar = 1$). This term is suitable because it has real expectation values. This follows from the condition of being Hermitian, so that $\langle \phi | i\frac{d}{dk}\phi \rangle = \overline{\langle \phi | i\frac{d}{dk}\phi \rangle}$ in the ordinary (nonrelativistic) inner product. Using nonrelativistic eigenstates of position in momentum representation, this is satisfied as follows:

$$(2\pi)^{-1}\int_{-\infty}^{\infty}\left(e^{ikx}\,i\frac{d}{dk}e^{-ikx}\right)dk = x\delta(0).$$

In the covariant inner product defined above, this is no longer the case. The second term of $\hat{\xi}$ (A.1) makes the necessary corrections for it to be Hermitian in the new inner product.

We may recall that the zero value eigenstate in k-space is just $\sqrt{w(k)e^{ik0}} = \sqrt{w(k)}$ so that by applying the operator above one gets:

$$\left(i\frac{d}{dx} - \frac{ik}{2w(k)^2}\right)\sqrt{w(k)} = i\frac{d}{dk}\left(k^2 - m^2\right)^{1/4} - \frac{ik}{2w(k)^{3/2}}$$

$$= i2k\frac{\left(k^2 - m^2\right)^{-3/4}}{4} - \frac{ik}{2\left(k^2 - m^2\right)^{3/4}} = 0.$$

Now recalling the boosted state in momentum space representation (for zero eigenvalue) one has $\sqrt{w'(k)e^{ik0}} = \sqrt{w'(k)}$. At this point, in order to get a reasonable expression for this term in terms of known functions of k, one can apply the small velocity approximation already used above (see equation [6.30]). This indicates that $\sqrt{w'(k)} \cong \sqrt{w(k) + vk} \cong \sqrt{w(k)} + \frac{1}{2}vk / \sqrt{w(k)}$. One next attempts to find an operator ξ such that if $\hat{\xi} = \xi + F(k)$, the result is a zero eigenvalue in the boosted operator:

$$\left(\hat{\xi} + F(k)\right)\left(\sqrt{w(k)} + \tfrac{1}{2}vk / \sqrt{w(k)}\right) = 0. \tag{A.2}$$

We first note that

$$\left(\hat{\xi}\right)\left(\sqrt{w(k)} + \tfrac{1}{2}vk / \sqrt{w(k)}\right) = \left(i\frac{d}{dk} - i\frac{k}{2w(k)^2}\right)\left(\sqrt{w(k)} + \tfrac{1}{2}vk / \sqrt{w(k)}\right)$$

$$= 0 + \left(\left(i\frac{d}{dk} - i\frac{k}{2w(k)^2}\right)\frac{1}{2}vk\left(w(k)\right)^{-1/2}\right)$$

$$= \frac{i}{2}v\left(w(k)\right)^{-1/2} - \frac{i}{2}k^2 v\left(w(k)\right)^{-5/2}$$

We may then look for an $F(k)$ so that the whole expression (A.2) gives a zero eigenvalue, i.e.,

$$\frac{i}{2}v\left(w(k)\right)^{-1} - \frac{i}{2}vk^2\left(w(k)\right)^{-3} + F(k)\left(1 + \frac{1}{2}vk\left(w(k)\right)^{-1}\right) = 0.$$

For small v, one concludes finally that

$$F(k) \cong -\frac{1}{2} i \frac{v}{(w(k))^3},$$

so that the confining operator on a slightly tilted hyperplane is

$$\hat{\xi} = \hat{\xi} - \frac{1}{2} i \frac{v}{(w(k))^3}.$$

We may now note that this operator is not Hermitian in the covariant inner product, so its eigenvectors are not orthogonal. This is the same situation one has for x on its own—of course $\hat{\xi}$ itself is Hermitian.

Appendix B

This derivation follows directly from equation (6.46) and is a solution of the triple integral expressed there. This derivation provides further detail to the case study provided in Chapter 6, to the end of arguing for an understanding of Hegerfeldt nonlocality as related to a failure of invariance, in the form of the Jericho effect.

Picking up the argument from (6.46) and recalling that here the primary interest is in the values of $\Psi(x,\varepsilon) = \Psi'(x',0)$ for large x and x', one notes that this is equivalent to the small velocity condition that $v/c \ll 1$. This implies that in this regime γ is approximately equal to 1. Indeed, using one of the Lorentz transformations above, one can conclude that

$$\begin{aligned} t' &= \gamma\left(t - vx / c^2\right) \\ \Rightarrow 0 &= \gamma\left(\varepsilon - vx / c^2\right) \\ \Rightarrow \frac{\varepsilon}{x} &= \frac{v}{c^2} \Rightarrow v = \frac{\varepsilon c^2}{x}. \end{aligned} \tag{B.1}$$

Thus the required velocity of the moving frame is proportional to ε and varies inversely as x.

Reintroducing the condition that $v/c \ll 1$, one finds that

$$\frac{\varepsilon c}{x} \ll 1, \tag{B.2}$$

which tells us the acceptable relationship between x and ε in order for a linear approximation to remain valid. As before, ε can be made arbitrarily small in order to use the approximation across the entire range of x values.

With these constraints one may expand in linear powers of v/c.

Using (B.1) and the Lorentz transformation for x' and setting units such that $c = 1$, one gets:

$$x' = \frac{1}{\sqrt{1 - \varepsilon^2/x^2}}\left(x - e^2/x\right)$$
$$\Rightarrow x' \cong x \tag{B.3}$$

which implies that $x' = x$ to order ε. Similarly, using the Lorentz transformation for momentum, one finds that

$$k = \frac{1}{\sqrt{1 - \varepsilon^2/x^2}}\left(k' + \frac{\varepsilon}{x}\, w(k')\right) \tag{B.4}$$

Since following from (B.1) $v = \varepsilon/x$, then keeping the linear term of v, this simplifies to

$$k = k' + \frac{\varepsilon}{x}\, w(k') \tag{B.5}$$

Looking at the Lorentz transform for $w(k)$ and using the same approximation, one gets:

$$w(k) = \frac{1}{\sqrt{1 - \varepsilon^2/x^2}}\left(w(k') + \frac{\varepsilon}{x}\, k'\right)$$
$$\Rightarrow w(k) \cong w(k') + \frac{\varepsilon}{x}\, k' \tag{B.6}$$

Now recalling that $w'(k') = w(k) = \sqrt{m^2 + k^2}$ one can rearrange (B.6) as follows:

$$w'(k') = w(k') + \frac{\varepsilon}{x}\, k'$$
$$\Rightarrow \frac{w'(k')}{w(k')} = 1 + \frac{\varepsilon/x\; k'}{w(k')} \tag{B.7}$$

Further simplifying, using the binomial theorem and keeping the linear term in ε, one gets:

$$\sqrt{\frac{w'(k')}{w(k')}} = 1 + \frac{1}{2}\frac{\varepsilon/_x\, k'}{w(k')}. \tag{B.8}$$

As one might expect, this is essentially the same expression found for the small velocity approximation (6.30). These expressions, (B.3) and (B.8), allow one to simplify the integrand of (6.46) to solve for probability amplitude $G(\xi)$ in this approximation, where ε is kept small compared to x.

Returning to the triple integral expression for $G(\xi)$ (6.46), a first step to evaluating this integral might be to change the order of integration. One can also use (B.3) to replace x' with x, which results in the following expression:

$$G(\xi) = (2\pi)^{-2} \int_{-\infty}^{\infty}\int_{-\infty}^{\infty}\int_{-\infty}^{\infty} e^{ik'x} \frac{w'(k')^{1/2}\, w(k)^{1/2}}{w(k')} e^{ik(\xi - x)} dk dk' dx. \tag{B.9}$$

The question of whether changing the order of integration is an acceptable method depends in part on the physical story that can be told here.

Proceeding with the approximation (B.8) above, then (B.9) becomes:

$$G(\xi) = (2\pi)^{-2} \int_{-\infty}^{\infty}\int_{-\infty}^{\infty}\int_{-\infty}^{\infty} e^{ik'x} \left(1 + \frac{\varepsilon}{2x}\frac{k'}{w(k')}\right) \frac{w(k)^{1/2}}{w(k')^{1/2}} e^{ik(\xi - x)} dk dk' dx$$

$$\Rightarrow G(\xi) = (2\pi)^{-2} \left(R + \int_{-\infty}^{\infty}\int_{-\infty}^{\infty} \frac{\varepsilon}{2x} \frac{e^{ik'x}\, k'}{w(k')^{3/2}} I dk dk' \right) \tag{B.10}$$

where $I = \int_{-\infty}^{\infty} w(k)^{1/2} e^{ik(\xi - x)} dk$ and $R = \int_{-\infty}^{\infty}\int_{-\infty}^{\infty}\int_{-\infty}^{\infty} e^{ik'x} \cdot e^{ik(\xi - x)} \frac{w(k)^{1/2}}{w(k')^{1/2}} dk dk' dx.$

In the integral involving the term I, the order of integration has been switched again to $dxdk'\,dk$. Setting the m term in $w(k)$ equal to one, the integral, I, becomes:

$$I = \int_{-\infty}^{\infty} \left(1 + k^2\right)^{1/4} e^{ik|\xi - x|} dk.$$

Replacing $|\xi - x|$ with x,

$$I = 2\int_0^\infty (1+k^2)\cos(kx)dk.$$

Again using Basset's formula, one notes that

$$\begin{aligned} I &= 2\left(1-\frac{d^2}{dx^2}\right)\int_0^\infty \frac{\cos(kx)}{(1+k^2)^{3/4}}dk \\ &= 2(1-\left(\frac{d^2}{dx^2}\right)x^{1/4}K_{1/4}(x)\frac{\Gamma(1/2)}{\Gamma(3/4)}\frac{1}{2^{1/4}}. \end{aligned} \tag{B.11}$$

Also noting that the $\Gamma(1/2)$ and $\Gamma(3/4)$ terms are just constants, one can rearrange (B.11), putting all the constants into a constant term, C, so that

$$\mathrm{I} = \mathrm{C}\left(1-\frac{d^2}{dx^2}\right)x^{1/4}K_{1/4}(x). \tag{B.12}$$

Now taking derivatives to evaluate (B.12), one gets:

$$\begin{aligned} I &= C\left[x^{1/4}K_{1/4}(x) - \frac{d^2}{dx^2}\left(x^{1/4}K_{1/4}(x)\right)\right] \\ &= C\left[x^{1/4}K_{1/4}(x) - \frac{d}{dx}\left(\frac{1}{4}x^{-3/4}K_{1/4}(x) + x^{1/4}K'_{1/4}(x)\right)\right] \\ &= C\left[x^{1/4}K_{1/4}(x) - \left(-\frac{3}{16}x^{-7/4}K_{1/4}(x) + \frac{1}{4}x^{-3/4}K'_{1/4}(x)\right.\right. \\ &\qquad \left.\left. + \frac{1}{4}x^{-3/4}K'_{1/4}(x) + x^{1/4}K''_{1/4}(x)\right)\right] \\ &= C\left[K_{1/4}(x)\left(x^{1/4} + \frac{3}{16}x^{-7/4}\right) - \frac{1}{2}x^{-3/4}K'_{1/4}(x) - x^{1/4}K''_{1/4}(x)\right] \end{aligned}$$

Using some standard relations for the K functions,

$$\begin{aligned} K_{-\nu} &= K_\nu \\ -2K'_\nu &= (K_{\nu+1} + K_{\nu-1}), \end{aligned}$$

one can derive the following expressions:

$$\begin{aligned} K'_{1/4} &= -\tfrac{1}{2}\left(K_{3/4} + K_{5/4}\right) \\ K''_{1/4} &= -\tfrac{1}{2}\left[-\tfrac{1}{2}\left(K_{1/4} + K_{7/4}\right) - \tfrac{1}{2}\left(K_{1/4} + K_{9/4}\right)\right] \\ &= \tfrac{1}{4}\left(2K_{1/4} + K_{7/4} + K_{9/4}\right). \end{aligned} \tag{B.13}$$

Plugging these back into the last expression for I, one discovers that

$$\begin{aligned} I = C\Bigg[& K_{1/4}\left(x^{1/4} + \frac{3}{16}x^{-7/4}\right) + \frac{1}{4}x^{-3/4}\left(K_{3/4} + K_{5/4}\right) \\ & - \frac{1}{4}x^{1/4}\left(2K_{1/4} + K_{7/4} + K_{9/4}\right)\Bigg] \end{aligned} \tag{B.14}$$

$$\Rightarrow I = \frac{C}{4}\left[K_{1/4}\left(2x^{1/4} + \frac{3}{4}x^{-7/4}\right) + x^{-3/4}K_{3/4} + x^{-3/4}K_{5/4} - x^{1/4}K_{7/4} - x^{1/4}K_{9/4}\right]$$

Re-substituting and expanding the constant term, C, then

$$\begin{aligned} I(\xi, x) = \frac{\Gamma(1/2)}{\Gamma(3/4)}\frac{1}{2^{5/4}}\Bigg[& K_{1/4}(|\xi - x|)\left(2|\xi - x|^{1/4} + \frac{3}{4}|\xi - x|^{-7/4}\right) \\ & + |\xi - x|^{-3/4}K_{3/4}(|\xi - x|) + |\xi - x|^{-3/4}K_{5/4}(|\xi - x|) \\ & - |\xi - x|^{1/4}K_{7/4}(|\xi - x|) - |\xi - x|^{1/4}K_{9/4}(|\xi - x|)\Bigg] \end{aligned} \tag{B.15}$$

Now evaluating the first term, R, in the integral expression (B.10) for amplitude $G(\xi)$,

$$(2\pi)^{-2}R = (2\pi)^{-2}\int_{-\infty}^{\infty}\int_{-\infty}^{\infty}\int_{-\infty}^{\infty} e^{ik'x}e^{ik(\xi - x)}\frac{w(k)^{1/2}}{w(k')^{1/2}}\,dk\,dk'\,dx$$

$$\Rightarrow (2\pi)^{-2}R = (2\pi)^{-1}\int_{-\infty}^{\infty}\int_{-\infty}^{\infty} e^{ik'\xi}\frac{w(k)^{1/2}}{w(k')^{1/2}}\delta(k' - k)\,dk'\,dk$$

$$\Rightarrow (2\pi)^{-2}R = (2\pi)^{-1}\int_{-\infty}^{\infty} e^{ik'\xi}\frac{w(k')^{1/2}}{w(k')^{1/2}}\,dk'$$

$$\Rightarrow (2\pi)^{-2}R = \delta(\xi)$$

At this point the original triple integral (B.9) can be rewritten in the form

$$G(\xi) = \delta(\xi)+(2\pi)^{-2}\int_{-\infty}^{\infty}\int_{-\infty}^{\infty}\frac{\varepsilon}{2x}\frac{e^{ik'x}k'}{w(k')^{3/2}}I(\xi,x)dxdk' \tag{B.16}$$

As above, Basset's formula may be applied to make the observation that

$$\int_0^\infty \frac{\cos(k'x)}{(1+k'^2)^{3/4}}dk' = |x|^{1/4}K_{1/4}(|x|)\frac{\Gamma(1/2)}{\Gamma(3/4)}\frac{1}{2^{1/4}}$$

$$\Rightarrow \int_0^\infty \frac{k'\sin(k'x)}{(1+k'^2)^{3/4}}dk' = -\frac{d}{dx}\left(|x|^{1/4}K_{1/4}(|x|)\right)\frac{\Gamma(1/2)}{\Gamma(3/4)}\frac{1}{2^{1/4}} \tag{B.17}$$

$$\Rightarrow \int_0^\infty \frac{k'\sin(k'x)}{(1+k'^2)^{3/4}}dk' = -S(x)\frac{d}{d|x|}\left(|x|^{1/4}K_{1/4}(|x|)\right)\frac{\Gamma(1/2)}{\Gamma(3/4)}\frac{1}{2^{1/4}}$$

where $S(x) = 1$, for $x > 0$, and $S(x) = -1$, for $x < 0$. So, simplifying (B.16) one gets:

$$G(\xi) = \delta(\xi) + (2\pi)^{-2}\, i\int_{-\infty}^{\infty}\int_0^{\infty}\frac{\varepsilon}{x}I(\xi,x)\frac{k'\sin(k'x)}{\left(1+k'^2\right)^{3/4}}dxdk'$$

$$\Rightarrow G(\xi) = \delta(\xi) + (2\pi)^{-2}\, i\frac{\Gamma(1/2)}{\Gamma(3/4)}\frac{1}{2^{1/4}}\int_{-\infty}^{\infty}\frac{\varepsilon}{x}I(\xi,x)S(x)\frac{d}{d|x|}\left(|x|^{1/4}K_{1/4}(|x|)\right)dx$$

$$\Rightarrow G(\xi) = \delta(\xi) + (2\pi)^{-2}\, i\frac{\Gamma(1/2)}{\Gamma(3/4)}\frac{1}{2^{1/4}}\int_{-\infty}^{\infty}-\frac{\varepsilon}{x}I(\xi,x)S(x)$$

$$\times\left(\frac{1}{4}|x|^{-3/4}K_{1/4}(|x|) + |x|^{1/4}K'_{1/4}(|x|)\right)dx$$

But, recalling that $K'_{1/4} = -\frac{1}{2}\left(K_{3/4} + K_{5/4}\right)$ from (B.13) above, one finally finds that

$$G(\xi) = \delta(\xi) + (2\pi)^{-2}\, i\frac{\Gamma(1/2)}{\Gamma(3/4)}\frac{1}{2^{1/4}}\int_{-\infty}^{\infty}-\frac{\varepsilon}{x}I(\xi,x)S(x)$$

$$\times\left(\frac{1}{4}|x|^{-3/4}K_{1/4}(|x|) - \frac{1}{2}|x|^{1/4}K_{3/4}(|x|) - \frac{1}{2}|x|^{1/4}K_{5/4}(|x|)\right)dx$$

Now actually performing the final x-integration, one confronts a difficulty caused by the inversion of the order of integration. Namely, the integrand diverges when evaluated at $x = \xi$. This is apparent as the various K_ν functions in I go to infinity at $x = \xi$. One can circumvent this by performing the x-integration with limits of integration $\pm L$, keeping $|\xi| > L$. In other words, one wants to pick up contributions to the amplitude $G(\xi)$ from the region where the exponential asymptotic form of the Hankel functions dominates. In order to do so one must consider the limit in which L goes to infinity but in which x goes to infinity after ξ.

Adopting this approach, then the expression for $G(\xi)$ has the final form, for a small time interval $t = \varepsilon$,

$$G(\xi) = \delta(\xi) + i\varepsilon f(\xi) e^{-|\xi|} \tag{B.18}$$

where

$$f(\xi) = \left(\int_{-L}^{-K\varepsilon} + \int_{K\varepsilon}^{L} \right) Q(x,\xi)\,dx. \tag{B.19}$$

The integration is carried out letting L go to infinity with $|\xi| > L$, so as not to pick up the (infinite) contribution to the integrand at $\xi = x$, and the function $Q(x, \xi)$ can be expanded as:

$$\begin{aligned} Q(x,\xi) = {} & \frac{\Gamma^2(\frac{1}{2})}{\Gamma^2(\frac{3}{4})} \frac{1}{2^{5/2}} (2\pi)^{-2} \frac{e^{|\xi|}}{x} S(x) \Big[K_{1/4}(|\xi - x|) \\ & \times \left(2|\xi - x|^{1/4} + \frac{3}{4}|\xi - x|^{-7/4} \right) + |\xi - x|^{-3/4} K_{3/4}(|\xi - x|) + (|\xi - x|)^{-3/4} \\ & K_{5/4}(|\xi - x|) - |\xi - x|^{1/4} K_{7/4}(|\xi - x|) - |\xi - x|^{1/4} K_{9/4}(|\xi - x|) \Big] \\ & \times \left[|x|^{1/4} K_{5/4}(|x|) + |x|^{1/4} K_{3/4}(|x|) - \frac{|x|^{-3/4}}{2} K_{1/4}(|x|) \right] \end{aligned} \tag{B.20}$$

Expression (B.19) for the integration over x excludes contributions from inside the light-cone, $x \in [-\varepsilon, \varepsilon]$, as well as from intervals $[-K\varepsilon, -\varepsilon]$ and $[\varepsilon, K\varepsilon]$ for $K \gg 1$. Inside the light-cone this approximation method does not apply. Since the integrand and the interval over which it would be taken, $[-\varepsilon, \varepsilon]$, are each of order ε the contribution to the integral would have been of order ε^2 and can be ignored in the linear approximation. The approximation also breaks down in the regions just outside the light-cone, intervals $[-K\varepsilon, -\varepsilon]$ and $[\varepsilon, K\varepsilon]$; this contribution to the integral is similarly of order $K\varepsilon^2$ and can also be ignored.

As expected, this result is also independent of the constant K. This is apparent as one considers that the asymptotic forms of the modified Hankel functions in Q are generally exponential, so that the x-integration in (B.19) will effectively have the form

$$\int_{K\varepsilon}^{L\to\infty} e^{-x}\,dx = e^{-K\varepsilon} \approx 1 - K\varepsilon + \ldots, \tag{B.21}$$

which for small values of ε is essentially independent of K.

In addition, the integral (B.19) will not depend exponentially on $|\xi|$. In performing the integration, one maintains $|\xi|$ as larger than L as it goes to infinity; for small x the first squared bracket term in Q (B.20) has the asymptotic behavior of an exponential function $e^{-|\xi|}$, which cancels the exponential factor in Q. For large x approaching ξ the second square bracket has the same exponential asymptotic behavior.

The form of this second result, (B.18), is basically that expected from the direct solution of the Hegerfeldt integral for small t, (6.40) above. The second method, while not as rigorous, suggests that the nonlocal spreading of the wave-packet can be seen as due to the Jericho effect. In other words, the two difficulties with Newton-Wigner localization, Hegerfeldt nonlocality (time evolution) and the Jericho effect (the noncovariance of localization on shifted hyperplanes), are essentially related phenomena.

Notes

Preface

1. For Lee Smolin, the role of broken symmetry is vitally important as he has expressed recently in his argument, "Against symmetry," in which he, a lone voice, explains what the authors would term a "relational strategy," but one controlled by dynamical laws (Lee Smolin, personal communication). For the case of elementary particle physics, broken symmetry is indeed very important. See Michael Redhead, *From Physics to Metaphysics* (Cambridge: Cambridge University Press, 1995), 66.

Introduction

1. Hermann Weyl, *Symmetry* (Princeton: Princeton University Press, 1982; originally published in 1952), 132.
2. David Malament, "Causal Theories of Time and the Conventionality of Simultaneity," *Nôus* 11, (1977): 293–300.

1. Scientific Representation

1. In this and subsequent schematics, the authors intend "Mind" to stand for the claims about the world that are generally accepted by the scientific community.
2. A. D. Smith, in his book of the same name, phrases the problem of perception as the "question of whether we can ever directly perceive the physical world." See A. D. Smith, *The Problem of Perception* (Cambridge, MA: Harvard University Press, 2002), 1.
3. The general approach of bracketing out the notions of "Mind" and "World" and focusing instead on media was partly inspired by Latour; see Bruno Latour, "How to be Iconophilic in Art, Science, and Religion?" in *Picturing Science Producing Art,* ed. Caroline A. Jones and Peter Galison (New York: Routledge, 1998), 426.
4. Wittgenstein considers that in using a private language a person merely *appears* to understand their own use of language, but only does so in an entirely sub-

jective manner. This subjectivity thus rules out there being a correct meaning associated with the private language. Applied to this account of scientific representation, this approach would suggest that at best a solitary representer might appear to be representing reality to themselves, but this again in an entirely subjective manner. See Ludwig Wittgenstein, *Philosophical Investigations,* I, 269, trans. G. E. M. Anscombe, 2nd ed. (Oxford: Blackwell, 1953), 94[e].

5. The analysis of formal relationships between structures only captures certain relevant aspects of scientific representation; a more complete account is provided in Chapter 2.
6. For a discussion of resemblance in visual art, see E. H. Gombrich, *Art and Illusion: A Study in the Psychology of Pictorial Representation* (Oxford: Phaidon Press, 1977, originally published in 1960), 29–78; for a discussion of the expression of artists' intentions (especially in the performance arts), see Malcolm Budd, *Music and the Emotions: The Philosophical Theories* (London: Routledge & Kegan Paul, 1985), 121–176.
7. See, for example, Nelson Goodman, *Languages of Art: An Approach to a Theory of Symbols* (Indianapolis: Hackett Publishing, 1976), 3–6.
8. According to John Searle, the concept of representation has simply been "abused," and Bas van Fraassen goes further to conclude that it is "too meager" to be of much use; see John R. Searle, *Intentionality: An Essay in the Philosophy of Mind* (Cambridge: Cambridge University Press, 1983), 11–13; see also Bas van Fraassen, "Interpretations of Science: Science as Interpretation," in *Physics and Our View of the World,* ed. Jan Hilgevoord (Cambridge: Cambridge University Press, 1994), 171–172.
9. In using these two distinctions, this follows Redhead; see Michael Redhead, "The Intelligibility of the Universe," in A. O'Hear, ed., *Philosophy at the New Millennium* (Cambridge: Cambridge University Press, 2001), 73–90.
10. Stewart Shapiro, *Philosophy of Mathematics: Structure and Ontology* (Oxford: Oxford University Press, 1997), 38; Michael D. Resnik, *Frege and the Philosophy of Mathematics* (Ithaca: Cornell University Press, 1980), 162.
11. Indeed there are two other possibilities: physical structures used to represent other physical structures and mathematical structures used to represent other mathematical ones.
12. Shapiro, *Philosophy of Mathematics,* 40–41.
13. This is adapted from Shapiro in the sense that he speaks in terms of sets of algebraic structures classified as such by the fact that they are not isomorphism classes, implying that it is possible to point to an individual algebraic structure. However, in the framework set up here, abstract structures may be classified as algebraic if the members of sets of concrete structures on which they are based are not isomorphic; it is therefore only possible to speak of algebraic *abstract* structures.
14. Shapiro, *Philosophy of Mathematics,* 41.
15. See Newton C. A. da Costa and Steven French, "The Model-Theoretic Ap-

proach in the Philosophy of Science," *Philosophy of Science* 57 (1990): 248–265; for further discussion, see also: Irene Mickenberg, Newton C. A. da Costa, and Rolando Chuaqui, "Pragmatic Truth and Approximation to Truth," *The Journal of Symbolic Logic* 51 (1986): 201–221; Otávio Bueno, "Empirical Adequacy: A Partial Structures Approach," *Studies in History and Philosophy of Science* 28 (1997): 585–610; Steven French and James Ladyman, "Superconductivity and Structures: Revisiting the London Account," *Studies in History and Philosophy of Modern Physics* 28 (1997): 363–393.

16. For a recent example, see Mauricio Suárez, "Scientific Representation: Against Similarity and Isomorphism," *International Studies in the Philosophy of Science* 17/3 (October 2003): 225–244; see also Suárez, "Theories, Models, and Representations," in *Model Based Reasoning in Scientific Discovery,* ed. Magnani, Nersessian, and Thagard (Dordrecht: Kluwer, 1999), 75–83; see also Roman Frigg, "Scientific Representation and the Semantic View of Theories," *Theoria* 55 (2006): 49–65.
17. Michael Redhead, "Symmetry in Intertheory Relations," *Synthese* 32 (1975): 77–112; see also Michael Redhead, "The Interpretation of Gauge Symmetry," in *Symmetries in Physics: Philosophical Reflections,* ed. Brading and Castellani (Cambridge: Cambridge University Press, 2003), 124–162; originally published in *Ontological Aspects of Quantum Field Theory,* ed. M. Kuhlmann, H. Lyre, and A. Wayne (River Edge, NJ: World Scientific, 2002), 281–301.
18. For a relatively recent collection of papers that attempt to do this, see Jones and Galison, eds., *Picturing Science Producing Art.*
19. Michael Polanyi suggests that, in using tools and instruments, "We accept them existentially by dwelling in them"; see Michael Polanyi, *Personal Knowledge: Towards a Post-Critical Philosophy* (London: Routledge, 1962, originally published in 1958), 59.
20. For one recent attempt, see Fred Muller, *Structures for Everyone,* Ph.D. diss., University of Utrecht, 1998.
21. It is worth pointing out that not all commentators support total freedom of interpretation within the philosophical discussion of representation in the arts. Richard Wollheim, for instance, rejects one concept of intention precisely because it leads to the total irrelevance of artistic intention for the viewer; see Richard Wollheim, *Art and Its Objects* (Cambridge: Cambridge University Press, 1980), 19–20.

2. Models, Symmetry, and Convention

1. Felix Klein, *Elementary Mathematics from an Advanced Standpoint: Geometry,* trans. E. R. Hedrick and C. A. Noble (New York: Dover, 1939; translated from 1925 edition), 69.
2. This way of thinking about transformations is exemplified within the branch of contemporary mathematics called "category theory"; see F. William Lawvere and Stephen H. Schanuel, "Part 1: The Category of Sets," in *Conceptual*

Mathematics: A First Introduction to Categories (Cambridge: Cambridge University Press, 1997), 13–36.

3. See, for example, Hermann Weyl, *The Classical Groups: Their Invariants and Representations* (Princeton: Princeton University Press, 1939).
4. For a discussion of the representation of a group, see J. P. Elliott and P. G. Dawber, *Symmetry in Physics,* vol. 1 (London: Macmillan, 1979), 43–84.
5. Klein, *Elementary Mathematics,* 58.
6. See, for instance, L. Fonda and G. C. Ghirardi, *Symmetry Principles in Quantum Physics* (New York: Marcel Dekker, 1970), 30–47.
7. Klein, *Elementary Mathematics,* 69.
8. Michael Redhead, "The Intelligibility of the Universe," in A. O'Hear, ed., *Philosophy at the New Millennium* (Cambridge: Cambridge University Press, 2001), 73–90.
9. Ibid.
10. Ibid.
11. For a historical treatment, see Roberto Torretti, *Philosophy of Geometry from Riemann to Poincaré* (Dordrecht: D. Reidel, 1978), 320–358.
12. It has been suggested that the term was coined in the German-speaking world, where Poincaré's conventionalism was generally understood as an alternative to Kant's geometric apriorism; see Jerzy Giedymin, *Science and Convention: Essays on Henri Poincaré's Philosophy of Science and the Conventionality Tradition* (Oxford: Pergamon Press, 1982), viii.
13. Torretti, *Philosophy of Geometry,* 320.
14. This connection has been noted recently by Giedymin, *Science and Convention,* vii.
15. Henri Poincaré, "Sur les hypothèses fondamentales de la géométrie," *Bulletin de la Société Mathématique de France* 15 (1887): 215; here quoted in translation by Torretti, *Philosophy of Geometry,* 335.
16. Ibid.
17. Henri Poincaré, *Science and Method,* trans. Francis Maitland (Bristol: Thoemmes Press, 1914), 100–114.
18. Hans Reichenbach, *The Philosophy of Space and Time,* trans. Maria Reichenbach and John Freund (New York: Dover, 1958), footnote 3, 36.
19. Ibid., 33.
20. In fact, Jerzy Giedymin has argued that Reichenbach, along with Moritz Schlick of the famous 'Vienna Circle' and others, misinterpreted Poincaré's philosophy precisely by failing to see the significance of group theoretical invariance within his conception of conventionalism; see Giedymin, *Science and Convention,* viii–ix.
21. Reichenbach, *Philosophy of Space and Time,* 35.
22. Ibid., 14–19.
23. Ibid., 37.
24. Ibid.
25. W. V. Quine, "Two Dogmas of Empiricism," in *From a Logical Point of View,*

2nd ed. (Cambridge, MA: Harvard University Press, 1961), 20–46; originally published in *Philosophical Review,* 60/1 (Jan. 1951): 20–43.

26. Ibid., 42.
27. Ibid., 44.

3. A New Appraisal of Symmetry

1. Brading and Castellani have also provided a useful survey of the different uses of symmetry in physics that includes the three themes addressed here, but does not discuss the sense in which they may be intended to work together; see Katherine Brading and Elena Castellani, "Introduction," in *Symmetries in Physics: Philosophical Reflections,* ed. Brading and Castellani (Cambridge: Cambridge University Press, 2003), 11–15.
2. Theories in Lagrangian form generate conservation laws from symmetries by Noether's theorem.
3. A. Einstein, "On the Electrodynamics of Moving Bodies," in *The Principle of Relativity,* trans. W. Perrett and G. B. Jeffery (New York: Dover, 1923), 35–65; originally published in *Annalen der Physik* 17 (1905).
4. Vincent Icke, *The Force of Symmetry* (Cambridge: Cambridge University Press, 1995), 103.
5. Steven Weinberg, *Dreams of a Final Theory* (London: Hutchinson, 1993), 109; 169.
6. Sunny Y. Auyang, *How Is Quantum Field Theory Possible?* (New York: Oxford University Press, 1995), 34.
7. Bas C. Van Fraassen, *Laws and Symmetry* (Oxford: Clarendon Press, 1989), 103.
8. Eugene P. Wigner, *Symmetries and Reflections* (Bloomington: Indiana University Press, 1967), 47.
9. Van Fraassen, *Laws and Symmetry,* 289.
10. Ibid.
11. Auyang, *How Is Quantum Field Theory Possible?,* 11.
12. Peter Kosso, "Symmetry, Objectivity, and Design," in Brading and Castellani, *Symmetries in Physics,* 413–419.
13. Hermann Weyl, *Symmetry* (Princeton: Princeton University Press, 1982; originally published in 1952), 132.
14. For a recent account of the different meanings of objectivity within the history of modern scientific practice, see Peter Galison, "Judgment Against Objectivity," in *Picturing Science Producing Art,* ed. Caroline A. Jones and Peter Galison (New York: Routledge, 1998), 327–359.
15. Similar approaches have been taken by Brown and by Nozick; this serves to confirm that these categories capture the major ways in which the term is used; see James Robert Brown, *Who Rules in Science?* (Cambridge, MA: Harvard University Press, 2001), 101–104; and Robert Nozick, *Invariances: The*

Structure of the Objective World (Cambridge, MA: Harvard University Press, 2001), 88.

16. The usual term, "inter-subjectivity," is sometimes taken to imply a universal agreement among all actual subjects; by using "multi-subjectivity" one may avoid this implication.
17. Bruno Latour, "How to Be Iconophilic in Art, Science, and Religion?" in *Picturing Science Producing Art,* ed. Caroline A. Jones and Peter Galison (New York: Routledge, 1998), 425.
18. Ibid., 426.
19. This result is also very similar to the form of sociological objectivity proposed by Bloor; see David Bloor, "A Sociological Theory of Objectivity," in *Objectivity and Cultural Divergence,* ed. S. C. Brown (Cambridge: Cambridge University Press, 1984), 229–245.
20. Weyl, *Symmetry,* 132.
21. It is not my intention here to address the distinct question of Weyl's philosophical perspective on objectivity and subjectivity as they appear in considering sense perception. The primary concern here is to understand his claim regarding the use of group theoretical invariance and models in physics, particularly as it represents a general trend among physicists and philosophers; it is in reference to this trend that we use the shorthand, "Weyl's approach to objectivity." On issues having to do with sense perception, see Hermann Weyl, *Mind and Nature* (Philadelphia: University of Pennsylvania Press, 1934).
22. Weyl, *Symmetry,* 128.
23. Ibid., 126.
24. Nozick, *Invariances,* 84.
25. Ibid.
26. Ibid., 106–111.
27. Ibid., 83 and 87–90; Nozick makes additional comments in an endnote that appear to take back these remarks—see note 21, pp. 332–333.
28. Ibid., 88.
29. Ibid., 90.
30. Ibid., 119.
31. "Accessibility from different angles" is in fact Robert Nozick's term to describe the hallmarks of objectivity; see Nozick, *Invariances,* 75–76.
32. These categories follow roughly those suggested by Redhead, "Symmetry in Intertheory Relations," *Synthese* 32 (1975): 77–112.
33. Ibid., 81.
34. Ibid., 106.
35. See Michael Redhead, "The Interpretation of Gauge Symmetry," in *Symmetries in Physics: Philosophical Reflections,* ed. Brading and Castellani (Cambridge: Cambridge University Press, 2003), 124–162; originally published in *Ontological Aspects of Quantum Field Theory,* eds. M. Kuhlmann, H. Lyre, and A. Wayne (World Scientific, 2002), 281–301.
36. Weyl, *Symmetry,* 132.

37. Ibid., 128.
38. For a discussion of the various forms of structural realism see James Ladyman, "What is Structural Realism?," *Studies in the History and Philosophy of Science* 29/3 (1998): 409–424.
39. Ibid., 422.
40. Michael Redhead, "The Intelligibility of the Universe," in *Philosophy at the New Millennium,* ed. A. O'Hear (Cambridge: Cambridge University Press, 2001), 73–90.
41. John Earman, *World Enough and Space-Time: Absolute versus Relational Theories of Space-Time* (Cambridge, MA: The MIT Press, 1989), 45–47.
42. Having introduced perspectival invariantism it is worth pointing out an aspect of its relationship to the account of scientific representation portrayed in Chapter 1. Since in this case the group of automorphisms may be interpreted in a perspectival sense, then individual agents within a social chain of media could each in principle occupy any one of the possible perspectives thereby allowed. Thus, each could verify (from their own perspective) as well as transmit the claim that *M* represents *O*. In this idealized sense, alignment of the social chain of media can be tied explicitly to the alignment of the formal chain of media, since perspectives are here dependent on the symmetries of *O* and *M*. In this sense, within the perspectival invariantist context, the alignment of the social and formal chains of media each utilize the mathematical apparatus of group theoretical invariance. This is not the case for invariantism more generally, and yet it may be that this fact contributes to the appeal of this attempt to construct a link between symmetry and objectivity.

4. Simultaneity and Convention

1. A. Einstein, "On the Electrodynamics of Moving Bodies," in *The Principle of Relativity,* trans. W. Perrett and G. B. Jeffery (New York: Dover, 1923), 35–65; originally published in *Annalen der Physik* 17 (1905).
2. These hyperplanes are so designated because they are the four-dimensional spacetime analogs of the usual three-dimensional notion of a plane.
3. Hans Reichenbach, *The Philosophy of Space and Time,* trans. Maria Reichenbach and John Freund (New York: Dover, 1958), 123.
4. On this see R. Anderson, I. Vetharaniam, and G. E. Stedman, "Conventionality of Synchronisation, Gauge Dependence and Test Theories of Relativity," *Physics Reports* 295 (1998): 95–97.
5. Reichenbach, *Philosophy of Space and Time,* 125–127.
6. Wesley C. Salmon, "The Philosophical Significance of the One-Way Speed of Light," *Noûs* 11 (1977): 271.
7. John A. Winnie, "Special Relativity without One-Way Velocity Assumptions: Part I," *Philosophy of Science* 37 (1970): 81–99; also see Carlo Giannoni, "Relativistic Mechanics and Electrodynamics without One-Way Velocity Assumptions," *Philosophy of Science* 45 (1978): 17–46.
8. Reichenbach, *Philosophy of Space and Time,* 143–147.

9. Ibid., 145.
10. Hermann Weyl, *Philosophy of Mathematics and Natural Science,* trans. Olaf Helmer (Princeton: Princeton University Press, 1949), 101–102.
11. David Malament, "Causal Theories of Time and the Conventionality of Simultaneity," *Noûs* 11 (1977): 293–300.
12. Sarkar and Stachel mention the "many advantages" of Einstein's criterion, this first among them; see Sahotra Sarkar and John Stachel, "Did Malament Prove the Non-Conventionality of Simultaneity in the Special Theory of Relativity?" *Philosophy of Science* 66/2 (1999): 209–220.
13. These include R. G. Spirtes, *Conventionalism and the Philosophy of Henri Poincaré,* Ph.D. diss., University of Pittsburgh, 1981; as well as Sarkar and Stachel, "Did Malament Prove the Non-Conventionality of Simultaneity?"
14. See A. A. Robb, *A Theory of Time and Space* (Cambridge: Cambridge University Press, 1914); as well as A. I. Janis, "Simultaneity and Conventionality," in *Physics, Philosophy and Psychoanalysis,* ed. R. S. Cohen and L. Laudan (Dordrecht: Reidel, 1983), 101–110; and Michael Redhead, "The Conventionality of Simultaneity," in *Philosophical Problems of Internal and External Worlds: Essays on the Philosophy of Adolf Grünbaum,* ed. J. Earman, A. I. Janis, G. J. Massey, and N. Rescher (Pittsburgh: University of Pittsburgh Press, 1993), 103–128.
15. The expression "local simultaneity" refers to something like the familiar scenario of synchronizing two or more spatially separated clocks, using light or radio signals. "Global simultaneity" refers to the mathematical partitioning, or foliation, of spacetime into hypersurfaces of constant time. Obviously, these two activities are closely related, but they each raise distinctive questions.
16. Reichenbach, *Philosophy of Space and Time,* 135–143.
17. For a discussion of Grünbaum's approach to the conventionality of simultaneity, see Redhead, "Conventionality of Simultaneity," 106–108.
18. Ibid., 15.
19. Salmon, "Philosophical Significance of the One-Way Speed of Light," 255.

5. Objectivity in the Twin Paradox

1. L. Marder, *Time and the Space Traveler* (Philadelphia: University of Pennsylvania Press, 1971), 91.
2. Paul Langevin, "L'évolution de l'espace et du temps," *Scienta* 10 (1911): 31–54.
3. H. Minkowski, "Space and Time," in *The Principle of Relativity,* trans. W. Perrett and G. B. Jeffery (New York: Dover, 1923), 73–91; translated from an address delivered at the 80th Assembly of German Natural Scientists and Physicians, Cologne, 21 September, 1908.
4. Langevin, "L'evolution de l'espace et du temps," 49.
5. The term "clock paradox" is sometimes also used to refer to the symmetry of time dilation between relatively moving frames, as contrasted with the twin

paradox, which refers to the asymmetry of elapsed proper time between the twins.

6. Marder, *Time and the Space Traveler,* 91.
7. J. C. Hafele and R. E. Keating, "Around the World Atomic Clocks: Predicted Relativistic Time Gains," *Science* 177 (1972): 166–168; "Around the World Atomic Clocks: Observed Relativistic Time Gains," *Science* 177 (1972): 168–170.
8. A. Einstein, "On the Electrodynamics of Moving Bodies," in *The Principle of Relativity,* trans. W. Perrett and G. B. Jeffery (New York: Dover, 1923), 43; originally published in *Annalen der Physik* 17 (1905).
9. Ibid.
10. Hans Reichenbach, *The Philosophy of Space and Time,* trans. Maria Reichenbach and John Freund (New York: Dover, 1958), 14–19.
11. Ibid., 15.
12. Lawrence Sklar, *Space, Time, and Spacetime* (Berkeley: University of California Press, 1974), 253–254; 256.
13. On dynamics within special relativity see Michael Friedman, *Foundations of Space-Time Theories: Relativistic Physics and the Philosophy of Science* (Princeton: Princeton University Press, 1983), 142–146.
14. Jan Hilgevoord and Joss Uffink have shown that, considered from within a quantum mechanical theoretical framework, these sorts of systems present characteristics that link them to an energy/time uncertainty relation and therefore to the Hamiltonian and thus to the dynamical concept of energy; see Jan Hilgevoord, "The Uncertainty Principle for Energy and Time: II," *American Journal of Physics* 66/5 (1998): 396–402.
15. See H. R. Brown, *Physical Relativity: Space-time Structure from a Dynamical Perspective* (Oxford: Clarendon Press, 2005).
16. The most famous dissenter was physicist Herbert Dingle. For a detailed historical treatment of his views see Hasok Chang, "A Misunderstood Rebellion: The Twin Paradox Controversy and Herbert Dingle's Vision of Science," *Studies in the History and Philosophy of Science* 24 (1993): 741–790.
17. Henri Arzelius, *Relativistic Kinematics* (Oxford: Pergamon, 1966), 189.
18. The authors have done just this; for our complete treatment see T. A. Debs and M. L. G. Redhead, "The Twin 'Paradox' and the Conventionality of Simultaneity," *American Journal of Physics,* 64/4 (1996): 384–392.
19. Langevin, "L'evolution de l'espace et du temps," 51–52.
20. David Bohm, *The Special Theory of Relativity* (New York: W. A. Benjamin, 1965), 168–170.
21. Ibid., 171.
22. Ibid.
23. For discussion of this version of the paradox see Wesley C. Salmon, *Space, Time, and Motion: A Philosophical Introduction* (Encino: Dickenson, 1975), 96–97; also see H. Bondi, "The Spacetraveller's Youth," *Discovery* 18 (1957): 505–510.

24. Salmon, *Space, Time, and Motion,* 97.
25. Bohm, *Special Theory of Relativity,* 166.
26. S. P. Boughn, "The Case of the Identically Accelerated Twins," *American Journal of Physics* 57 (1989): 791–799.
27. For discussion of this point, see in particular Edward A. Desloge and R. J. Philpott, "Uniformly Accelerated Reference Frames in Special Relativity," *American Journal of Physics* 55 (1987): 252–261.
28. Michael Redhead, "The Conventionality of Simultaneity," in *Philosophical Problems of Internal and External Worlds: Essays on the Philosophy of Adolf Grünbaum,* ed. J. Earman, A. I. Janis, G. J. Massey, and N. Rescher (Pittsburgh: University of Pittsburgh, 1993), 103–128.
29. John A. Winnie, "Special Relativity without One-Way Velocity Assumptions: Part I," *Philosophy of Science* 37 (1970): 81–99; see also Hanock Ben-Yami, "Causality and Temporal Order in Special Relativity," *British Journal for the Philosophy of Science* 57 (2006): 459–479.
30. Bohm, *Special Theory of Relativity,* 168–169.
31. Ibid., 170–171.
32. Ibid., 171.
33. Tevian Dray, "The Twin Paradox Revisited," *American Journal of Physics* 58 (1990): 822–825; R. J. Low, "An Acceleration-Free Version of the Clock Paradox," *European Journal of Physics* 11 (1990): 25–27.
34. J. C. Hafele and Richard E. Keating, "Around the World Atomic Clocks: Predicted Relativistic Time Gains," *Science* 177 (1972): 166–168; J. C. Hafele and Richard E. Keating, "Around the World Atomic Clocks: Observed Relativistic Time Gains," *Science* 177 (1972): 168–170.
35. Clive Kilmister and Barrie Tonkinson, "Pragmatic Circles in Relativistic Time Keeping," in *Correspondence, Invariance and Heuristics: Essays in Honor of Heinz Post,* ed. Steven French and Harmke Kamminga (Dordrecht: Kluwer Academic, 1993), 214.
36. Hermann Bondi, *Relativity and Common Sense: A New Approach to Einstein* (London: Heinemann, 1965), 34–35.
37. Boughn, "The Case of the Identically Accelerated Twins"; Edward A. Desloge and R. J. Philpott, "Comment on 'The Case of the Identically Accelerated Twins,' by S. P. Boughn," *American Journal of Physics* 59 (1991): 280–281.
38. See, for example, Barry R. Holstein and Arthur R. Swift, "The Relativity Twins in Free Fall," *American Journal of Physics* 40 (1972): 746–750.

6. Localization in Quantum Theory

1. An earlier version of some of this material appeared as Talal A. Debs and Michael L. G. Redhead, "The 'Jericho Effect' and Hegerfeldt Non-Locality," *Studies in the History and Philosophy of Modern Physics* 34 (2003): 61–85.
2. Jan Hilgevoord, "The Uncertainty Principle for Energy and Time," *American Journal of Physics,* 64/12 (1996): 1451–1456; and Jan Hilgevoord, "The Un-

certainty Principle for Energy and Time. II," *American Journal of Physics* 66/5 (1998): 396–402.

3. A. Einstein, B. Podolsky, and N. Rosen, "Can Quantum-Mechanical Description of Physical Reality be Considered Complete?" *Physical Review* 47 (1935): 777–780; J. S. Bell, "On the Einstein-Podolsky-Rosen Paradox," *Physics* 1 (1964): 195–200.
4. Michael Redhead, *Incompleteness, Nonlocality, and Realism: A Prolegomenon to the Philosophy of Quantum Mechanics,* 2nd rev. ed. (Oxford: Clarendon, 1989), 169.
5. Ibid., 141.
6. This standard version of the EPR paradox is originally due to D. Bohm, *Quantum Theory* (Englewood Cliffs, NJ: Prentice-Hall, 1951) 614–623.
7. Redhead, *Incompleteness, Nonlocality, and Realism,* 169.
8. P. Mittelstaedt, "Can EPR-correlations be used for the transmission of superluminal signals?" *Annalen der Physik* 7 (1998): 710.
9. Redhead, *Incompleteness, Nonlocality, and Realism,* 71.
10. Abner Shimony, *Search for a Naturalistic World View: Scientific Method and Epistemology,* vol. 1 (Cambridge: Cambridge University Press, 1993), 66.
11. Mittelstaedt, "Can EPR-correlations be used for the transmission of superluminal signals?" 710.
12. M. L. G. Redhead and Patrick La Riviere, "The Relativistic EPR Argument," in *Potentiality, Entanglement and Passion-at-a-Distance,* ed. R. S. Cohen et al. (Dordrecht: Kluwer, 1997), 207–215; G. Ghirardi and R. Grassi, "Outcome Predictions and Property Attribution: The EPR Argument Reconsidered," *Studies in the History and Philosophy of Science* 25 (1994): 397–423.
13. See David Malament, "In Defense of Dogma: Why There Cannot Be a Relativistic Quantum Mechanics of (Localizable) Particles," in *Perspectives on Quantum Reality: Non-Relativistic, Relativistic, and Field Theoretic,* ed. Rob Clifton (Dordrecht: Kluwer, 1996), 1–10.
14. Gerhard C. Hegerfeldt, "Remark on Causality and Particle Localization," *Physical Review D* 10/10 (1974): 3321.
15. Ibid., 3320.
16. M. L. G. Redhead, unpublished lecture, Cambridge, 1997.
17. M. J. Lighthill, *An Introduction to Fourier Analysis and Generalised Functions* (Cambridge: Cambridge University Press, 1958), 10–14.
18. See P. A. M. Dirac, *The Principles of Quantum Mechanics,* 4th ed. (Oxford: Clarendon, 1958, originally published in 1930), 72–76.
19. Hilgevoord, "The Uncertainty Principle for Energy and Time. II," 396.
20. Basset's formula is given as:

$$K_\nu(xz) = \frac{\Gamma\left(\nu + \frac{1}{2}\right)(2z)^\nu}{x^\nu \Gamma(\frac{1}{2})} \int_0^\infty \frac{\cos(xu)}{\left(u^2 + z^2\right)^{\nu+\frac{1}{2}}} du$$

where $(\nu + \frac{1}{2}) \geq 0$ and $x > 0$; see G. N. Watson, *Treatise on the Theory of Bessel Functions,* 2nd ed. (Cambridge: Cambridge University Press, 1922), 172.

21. The asymptotic form for the modified Hankel function $K_\nu(z)$ is given as:

$$K_\nu(z) \approx \left(\frac{\pi}{2z}\right)^{1/2} e^{-z}\left[1 + \frac{4\nu^2 - 1^2}{1!8z} + \frac{(4\nu^2 - 1^2)(4\nu^2 - 3^2)}{2!(8z)^2} + \ldots\right]$$

For our purposes the important feature is the exponential of the argument z; see ibid., 202.

22. Here the notation used is such that, for multiple integrals like this one, the order of integration is x first and then k.
23. The Bible, Book of Joshua, Chapter 6.
24. To further convince the skeptic about the implications of the Jericho effect, another way of expressing it is to think in terms of position operators; see Appendix A.
25. Hegerfeldt makes this point by noting the root in the exponent of the integrand; see Hegerfeldt, "Remark on Causality and Particle Localization," 3320; as well as G. C. Hegerfeldt, "Instantaneous Spreading and Einstein Causality in Quantum Theory," *Annalen der Physik* 7 (1998): 717.
26. Watson, *Treatise on the Theory of Bessel Functions,* 79.
27. For the full detail of this calculation, see Appendix B.
28. This example illustrates the meshing of dynamics and kinematics in relativistic quantum theory, since the Jericho effect is based on kinematics and Hegerfeldt's theorem on dynamical evolution of the state; one can also see a de-coupling of kinematics from dynamics in the nonrelativistic limit as the hyperplane geometry disappears.
29. Malament, "In Defense of Dogma," 1–10.
30. G. N. Fleming, "A Manifestly Covariant Description of Arbitrary Dynamical Variables in Relativistic Quantum Mechanics," *Journal of Mathematical Physics* 7 (1966): 1959–1981; G. N. Fleming and H. Bennett, "Hyperplane Dependence in Relativistic Quantum Mechanics," *Foundations of Physics* 13 (1989): 231–267.

Conclusion

1. Hermann Weyl, *Symmetry* (Princeton: Princeton University Press, 1982; originally published in 1952), 132.
2. Robert Nozick, *Anarchy, State and Utopia* (Basic Books, 1974), preface, xiii.
3. Ibid.

Index

Harvard University Press is a member of Green Press Initiative (greenpressinitiative.org), a nonprofit organization working to help publishers and printers increase their use of recycled paper and decrease their use of fiber derived from endangered forests. This book was printed on 100% recycled paper containing 50% post-consumer waste and processed chlorine free.